AI & I

AI & I

An Intellectual History of Artificial Intelligence

Eugene Charniak

foreword by Michael L. Littman

The MIT Press
Cambridge, Massachusetts
London, England

The MIT Press would like to thank the anonymous peer reviewers who provided comments on drafts of this book. The generous work of academic experts is essential for establishing the authority and quality of our publications. We acknowledge with gratitude the contributions of these otherwise uncredited readers.

This book was set in LATEX by Westchester Publishing Services. Printed and bound in the United States of America.

Library of Congress Cataloging-in-Publication Data

Names: Charniak, Eugene, author.
Title: AI & I : an intellectual history of artificial intelligence /
   Eugene Charniak ; foreword by Michael L. Littman.
Other titles: AI and I
Description: Cambridge, Massachusetts : The MIT Press, [2024] |
   Includes bibliographical references and index.
Identifiers: LCCN 2023058948 (print) | LCCN 2023058949 (ebook) |
   ISBN 9780262548731 (paperback) | ISBN 9780262379243 (epub) |
   ISBN 9780262379250 (pdf)
Subjects: LCSH: Artificial intelligence—History.
Classification: LCC Q335 .C48298 2024 (print) | LCC Q335 (ebook) |
   DDC 006.309—dc23/eng/20240224
LC record available at https://lccn.loc.gov/2023058948
LC ebook record available at https://lccn.loc.gov/2023058949

10   9   8   7   6   5   4   3   2   1

publication supported by a grant from
The Community Foundation for Greater New Haven
as part of the Urban Haven Project

To my colleagues. AI is *hard*.

The MIT Press gratefully acknowledges Michael Littman's work to help finalize Professor Charniak's manuscript after his untimely death in 2023.

Contents

Foreword

I visited Professor Eugene Charniak's office as a prospective graduate student in 1992. I had spent the previous four years as a junior member of a research group that steadfastly avoided the term "artificial intelligence"—after all, the cold breezes of the second AI winter were still blowing. Instead, my group had been focused on *statistical natural language processing*, in which the meanings of words arose, not from artisanally crafted data structures but from their observed colocations across thousands or even tens of thousands of appearances in messy, naturally occurring text. I was meeting with Eugene Charniak, a recognized leader in natural language processing, to ask whether, if I decided to come to Brown, he'd be interested in collaborating with me on the topic.

He more or less told me no. He preferred to work on *probabilistic* natural language processing, where the hand-designed representations of meaning that dominated the early days of the field were not jettisoned but instead were augmented by numeric information that helped focus the systems on *likely* interpretations instead of just plausible ones.

I hadn't heard this distinction before and, frankly, I was a little skeptical. I didn't see a path where either probabilistic or statistical natural language processing would bring about generally artificially intelligent systems. At least the statistical approach skirted the human bottleneck and held the promise of scaling up to real-world text.

I left the meeting excited that I had had a stimulating discussion with such a distinguished AI researcher. I didn't feel talked down to or dismissed but welcomed into a broader discussion where some very deep and very exciting problems were being considered. Reading this book, I'm sure you'll get a sense of what it was like discussing these topics with him. He had grappled with the challenge of making computers into capable language users from almost every conceivable angle. He had formed some extremely well-grounded opinions, but he was still open to consider a new angle if it was unfamiliar to him. He had a way of embodying the attitude of "Interesting! Let's think that through together." It made him a phenomenal contributor to the field, an inspiring professor, and an intellectual role model for many of us. Maybe "AI" wasn't such a shameful word after all.

I ended up joining the department at Brown as a graduate student a year later. My research was focused on the field of reinforcement learning, but there was some fascinating shared structure between language and decision-making—after all, both are centrally concerned with *sequences*, words in one and actions in the other, and how to generate them. I decided to check in on Eugene and was dumbfounded to discover that his research was now firmly rooted in statistical natural language processing. Moreover, he was teaching a class on the topic and was in the process of writing a textbook to help students appreciate the foundational concepts in the field. He had transformed himself from skeptic to world expert seemingly overnight, because he came to see statistical methods as the more promising path to writing programs that could use language effectively. In time, his work on statistical parsing—annotating a sequence of words with their relationships so the deeper meaning could be recognized—became the best in the world. With his students, he used statistical learning to build software that could "read" sentences from a collection of newspaper articles and parse them at a level that was nearly competitive with human annotators.

I ended up graduating from Brown with my PhD—the last class I ever took for a grade used Eugene's textbook! I returned about twenty years later as a faculty member. As a result, I was Eugene's colleague when the natural language processing field underwent its most recent transition, from statistical natural language processing to *neural* language models. As he had done with the transition to statistical natural language processing, Eugene began to teach a course on deep learning and wrote a textbook. He fully embraced this new perspective and its promise for solving problems he had wrestled with for decades, if not his whole career. His last published paper, "Parsing as Language Modeling," written with his last student, Do Kook Choe (DK), showed how his beloved parsing problem could be combined with advances in neural network training. Ultimately, DK's dissertation showed how to achieve human-level performance when parsing the collection of newspaper articles, allowing Eugene and DK to claim that their work had, once and for all, solved this problem.

Eugene had never seen parsing as the end goal of AI, but he definitely believed it was an important stepping-stone. He had chased the problem from logical, to probabilistic, to statistical, and, finally, to neural approaches, where, at last, it yielded. His thesis in this book is that artificial intelligence can now get started. I think you'll enjoy reading how he makes his case, as he shares his wit and insight from his front-row seat to over seventy years of work.

With AI achieving liftoff, I will speak on behalf of my colleagues in the field: Thank you, Eugene, for your role in getting us off the launch pad. Now the sky's the limit!

Michael L. Littman, January 2024

Preface

The field of artificial intelligence (AI) is about sixty-five years old, but already it plays an outsized role in the intellectual ferment of our age. It is *a* (perhaps *the*) major contributor to our current theory of mind and with good reason. It is also cited as a critical technology for our continued prosperity, though I think this latter claim is somewhat over hyped. On the other hand, AI's importance here is routinely paired with that of quantum computing, which is *hugely* overblown. (But that is a different book.)

This book is my stab at writing AI's intellectual history as a coherent whole. I say "intellectual history" because it is the story of the major ideas that have undergirded the discipline. A plain "history of AI" would give more dates and people, be encyclopedic, cover topics such as major organizations (the largest is the Association for the Advancement of AI, aka AAAI), funding, or its place in computer science (until recently, both low), and give biographies of the major contributors to the field.

An intellectual history of AI is of interest in its own right. This history makes the case that AI from the start had the right goal—writing computer programs that demonstrate intelligent behavior—but was completely wrong about how to achieve it. Specifically, we had two major misconceptions: first, that learning is a problem rather than a solution. If back at the beginning you asked an AI researcher what the major problem areas in the field were, they would have listed computer vision, inference, knowledge representation, understanding language, ... and learning. You would no more set out to translate between languages by writing a program to learn how to do so than you would create a rocket that learned to go to the moon. When put that way, it seems like common sense, but it is wrong.

Our second big mistake was our allegiance to the physical symbol hypothesis:

> A physical symbol system has the necessary and sufficient means for general intelligent action.

Here a "symbol" is a mark that represents something else, like a musical note representing a specific sound or a word representing an idea or object. A "physical symbol system" is a computer that manipulates symbols. Back in the day, when allegiance to this belief was near universal, it was credited to Allen Newell and Herbert Simon, who led AI research at Carnegie Mellon University. Indeed, Newell and Simon were the first to state this explicitly, but since in my narration this is cast as, well, a big mistake, let me hasten to add that virtually *everyone* believed it at that time—certainly your author.

The new gospel is well expressed by Geoffrey Hinton, who in my telling of this story is cast in the role of godfather of the new deep-learning-based AI:

> We now think of internal representation as great big vectors, and we do not think of logic as the paradigm for how to get things to work. We just think you can have these great big neural nets that learn, and so, instead of programming, you are just going to get them to learn everything. (https://www.forbes.com/sites/peterhigh/2016/06/20/deep -learning-pioneer-geoff-hinton-helps-shape-googles-drive-to -put-ai-everywhere/?sh=9ff6e35693c2)

Which brings us to the second reason—it is a good time for an intellectual history of AI. Your author fully subscribes to the new views. And while I believe that the logic of how we got to our current state is a powerful argument in its favor, this view is not universally accepted. Indeed, the publisher of this book, MIT Press, which published the first major deep-learning text [30] has published at least three subsequent books that argue that the new AI is overemphasized at best and misguided at worst [7, 56, 59].

Returning to this text, the title, *AI & I*, is a signal to the reader that this is an idiosyncratic account. I have spent my intellectual life as an AI practitioner, and it is my life in the field that has led me to the conclusions I have reached. However, I should warn that I am not a historian of science, and I do not have the global vision of the field that a scientific historian would possess before undertaking a definitive history. Rather my qualifications are my age, which has made for a long path through AI, and my ability to put down words to the end of a sentence or sentences to the end of a paragraph.

Because it is idiosyncratic, my account has definite shortcomings, the largest of which is I have included so little of the interesting ideas my colleagues have discovered. To have done otherwise—to have synthesized everything from AI—would have been the work of another lifetime. My dedication is an unsatisfactory substitute for the huge totality of these omissions. This book is also US centric as I have observed this history from my perches in the United States. My apologies to the French, Japanese, British, and others.

This book is aimed at the scientifically educated layperson. However, in many cases I cover ideas in more detail than such books are wont. I do so, in part, because I can, or at least I hope I can. The history of AI to this date, unfortunately, is one of researchers searching for a hold on a very difficult problem: how to construct an intelligent artifact out of non-intelligent pieces. It is only in the last fifteen years or so that we have been making anything like cumulative progress, and by going into detail I want to convince you that we AI researchers have not simply been wasting our time. The ideas we have had to reject, or at least set aside, have been reasonable, but they have not been productive. But to see this, and thus to better appreciate our current, much happier situation, it is necessary to see how things previously went wrong. For this I need the extra detail. If you do not find the result clear, let me know and I will try harder next time.

While I have you here, I should give some clues about organization. As a history should, I start at the beginning, and the last chapters are about the end. The middle is less clear. When a field has only a few practitioners, it is going to be unified if only because, even if there are recognizable subdisciplines, one has so few colleagues to look to. The unification at the end is a true convergence of areas due to the remarkable applicability of neural networks. However, quite soon in the history of AI, the subdisciplines diverged, and, say, folks in computer vision had quite little in common with those interested in games. Thus starting in chapter 2, chapters are single subdiscipline, and temporal order is most found within chapters. There are some temporal reasons for the order. For example, while computer vision research was establishing itself in the mid-1970s, it is not until the 1980s that it developed a unique style (at least in my mind). Thus readers will often find themselves transported back in time at the beginnings of chapters. To make this clearer, I have taken the advice of one reviewer and added dates to the chapter titles.

A few of my colleagues have given me helpful aid and advice. Ernie Davis shared his remarkable general knowledge of many parts of our field. Peter Norvig was the first person after myself to read the complete manuscript, and, by finding in it the book I set out to write, gave me a great deal of encouragement. Rao Kambhampati helped me see AI from the viewpoint of researchers in the AI subdiscipline of planning, though he will probably not agree with my take on things. And finally, my student David McClosky gave the book a remarkably close reading and helped me erase the evidence of my inability to see typos.

I am sure that were it not my intention to enforce a unity of outlook by not asking for help, the above list would be much longer, and perhaps the book much better. But it would not have been this book.

Providence, Rhode Island
May 2023

1

Beginnings (1956–1970)

1.1 Coming to AI

I was not around at the creation. The term "artificial intelligence" (AI) was coined in 1956 as the topic for a small academic workshop held at Dartmouth University, where AI was defined as research that followed from

> the conjecture that every aspect of learning or any other feature of intelligence can in principle be so precisely described that a machine can be made to simulate it [64].

That this is even *true*, much less achievable, was definitely a minority opinion in 1956. So, it behooves us to mention a few of the distinguished thinkers who made this a comfortable position for a few advanced computer scientists of this era. Foremost was the philosopher Thomas Hobbes, who in 1651 published his most important book, *The Leviathan* [37]. In it he states that "thinking is but reconning." His work is primarily a book of political philosophy. Its most famous quote concerns the life of primitive man, which he claimed was "solitary, poor, nasty, brutish, and short." By this Hobbes was saying that it is only by establishing governments over us that we escape this fate. His comments on thinking come in the introductory sections, where he grounds his philosophy of government in his ideas on human psychology. As for the term "reconning," the philosopher and mathematician Blaise Pascal had created a machine for addition, subtraction, and multiplication just a few years earlier, so it is reasonable to believe that it was these sort of operations that Hobbes had in mind when he used the term.

The other major contributor who set the intellectual agenda for the Dartmouth Conference was Alan Turing, who has as good a claim as anyone to be the father of all computer science and is also a hero of mine. In 1950, he published "Computing Machinery and Intelligence" [103], in which he explicitly states that the then emerging electronic computers were the right medium for the physical realization of an artificial intelligence. Thus, although Turing never used the term "artificial intelligence," 1950 is the second most popular date for the birth of the field. A good source for those interested in this period of early AI is Pamela McCorduck's *Machines Who Think* [65]. It is particularly noteworthy for McCorduck's interviews with the pioneers in the field.

I did not show up until 1968, the year I enrolled at MIT (the Massachusetts Institute of Technology) to study computer science (CS). Since there was no department of computer science—it was bundled inside electrical engineering—I sent my application to engineering but only after checking that I need not actually know any electrical engineering to fulfill the PhD requirements.

At that time, I was an undergraduate in physics at the University of Chicago. My knowledge of computer science stemmed from having taken a FORTRAN programming course at the end of my junior year. (To give you some idea of how far computers were from mainstream academia, the course was taught in the business school.) Then, during the summer between my junior and senior years, I got a job at Argonne National Laboratory with a group doing experiments to measure properties of neutrinos, a fundamental particle that is notoriously difficult to detect. It turned out that they needed a program to help decide when the atomic collision photographs they produced actually showed a neutrino so they could throw out the vast majority that did not. A physicist would then look at the good ones. When they learned I knew how to program, they assigned me to the effort.

(A brief aside: Programing was a much more energetic activity in those days. First, programs were typed onto lightweight cardboard punch cards about 3×8 inches in size with one instruction per card. However, the punch-card typewriter was in a different building. After creating all the cards, you transferred the program to a paper tape—a roll of about one-inch-wide black paper with the instructions encoded with small circular holes, or the lack thereof. Then you walked back to the physics lab, where your dedicated computer was located, and fed the tape into the paper-tape reader that was connected to the actual computer. If, as was always the case, the program had a bug, you figured out what was wrong and headed back to the building with the punch-card writer. Occasionally I was able to modify the paper tape to correct the program by punching an extra hole directly into the tape. I still am amazed that I could actually do that!)

At the end of the summer, I returned to University of Chicago for my senior year in physics, still intending to do graduate work and get my PhD in the field. How this changed in the course of the next six months is another story of how serendipity rules our lives.

The Chicago physics department ran an Undergraduate Journal Club where senior undergraduate physics students would give talks to the group about recent developments in the field. The scuttlebutt among undergraduates was that if you attended and did a presentation, the professor who ran the club would give you a great recommendation to grad school. So I joined. Unfortunately, I was not anywhere close to being able to read real physics articles about current physics (the first course in quantum mechanics was first semester senior year) so I agreed to talk about what I did over the summer: programing at Argonne Lab.

I went to the science library and took out every book I could find on computers. This came to about a linear foot. But one of them was a sort of "the year's best articles in computer science," and in it was an article about a checkers-playing program by a fellow at IBM named Arthur Samuel [91]. Samuel's program was written in 1952, so it has good credentials to be the first AI program ever written. Furthermore, it was certainly my entry to the field, so let's look at it in more detail.

1.2   Samuel's Checkers-Playing Program

Checkers is a game played on an 8×8 board of alternating red and black squares. One player controls the red pieces, the other black. All pieces start and stay on black squares and the players alternate moves. If your move causes you to land on a square containing one of your opponent's pieces, you capture it. When players have no possible move, they lose. (This is usually because all of their pieces have been captured.)

The key idea I want to introduce is that checkers, like so many AI problems, is one of *search*. Think of the checkerboard, plus pieces, plus an indicator showing who has the next move, as defining the complete state of the game. When a player makes a move, we go to the next state of the game. We can imagine a diagram with all possible states, starting with the initial positions of the pieces, with all states connected to their successor states by lines. This would look like a tree, with the initial state being the root of the tree and then quickly branching out because many possible next moves, and thus next states, exist.

While checkers is a pretty simple game compared to chess, for example, let's illustrate this with an even simpler game, tic-tac-toe. Figure 1.1 shows the root position and several follow-on positions for the tic-tac-toe game tree. The root, or starting position, is at the top (for some reason, most trees in AI are written this way, meaning the "tree" is upside down)

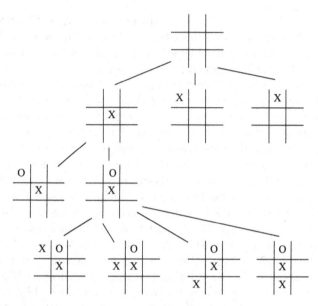

Figure 1.1: Root and a few possible next states for a tic-tac-toe game tree

and a few possible moves leading to different new states below. As always we assume the person playing X moves first.

Next we assign a score to all possible states. If X wins, the score is, say, 100; if O wins, −100; if neither player has won, 0. The goal for X is to get to a state with the maximum possible score, and for O, the minimum. This is referred to as *minimax search*. In the ideal case, the X player would search though all possible moves and countermoves. (And, in particular, X should assume that the opponent always chooses the move leading to the minimum score, that is, −100.) For tic-tac-toe some trial and error will convince you that if both players play optimally the game ends in a draw. (When I was a child, my father would take me to some cultural or educational activity on Saturday mornings. My favorite was the Museum of Science and Industry, where one of the exhibits was a machine that played optimal tic-tac-toe. Since I could never win, but I could draw, I quickly figured this out.)

We now know that optimal checkers also leads to a draw, but since this requires in effect exploring all possible checkers games, the computer power required to show this was way beyond anything we could muster back in 1952. So instead, Samuel's program depends on a *heuristic state evaluation function*. For example, in general if you have more pieces on the board than your opponent, you are more likely to win. If neither red nor black has won, you could assign a score to a state by counting the number of black pieces and subtracting the number of red. More generally, a *heuristic* is a rule of thumb that often works, but not always.

I was captivated by the idea of a program playing board games, but as a true physicist, writing a program to play checkers seemed like intellectual fluff compared to physics. So, when I applied to graduate school that fall, all my applications went to physics departments, not computer science. However, when I applied to the National Science Foundation for a fellowship (in physics), I said that perhaps I wanted to do some sort of physics that overlapped with computer science.

My appreciation of Samuel's program, however, led me to propose a second talk to the Journal Club, this time focused on artificial intelligence. In the course of preparation, i came upon the September 1966 issue of *Scientific American*, which was a special issue on information. In it was an article by Marvin Minsky titled "Artificial Intelligence" [71]. Also, the introductory article was by John McCarthy, another AI pioneer. I began to sense that perhaps computer science (and AI) was not fluff after all. I asked around and was told that Stanford and MIT were the best places for CS. I then asked which had the better off-campus life, Boston or Palo Alto. When I was told Boston, I applied to MIT to study computer science.

By that time I was very late in sending my application, but having received an NSF fellowship, MIT was happy to accept me anyway. I then asked NSF if I could use my fellowship for CS rather than physics with a side of CS. They agreed, and my life's work was settled.

1.3   AI around 1968

At the point when I came to the field, artificial intelligence was intellectually quite like the discipline born twelve years earlier. The majority of research was taking place at three academic centers, each represented by one or two attendees at the Dartmouth Conference: Marvin Minsky at MIT, John McCarthy at Stanford, and Allen Newell and Herbert Simon at Carnegie Mellon University.

While each of the three centers had its own intellectual flavor, the basic orientations of all three were quite similar. In retrospect, I would now describe the research primarily as symbolic, antinumeric, and topic-specific. By topic-specific I simply mean that a researcher would pick a topic—checkers, in the case of Samuel—or separating blocks in a visual scene, an example I take up shortly. This task-specific focus was pretty much a necessity since there was no understanding of what the fundamental problems might be that would span multiple problem areas. The tasks were chosen according to the researcher's taste but typically to fit the antinumeric biases of the time. I will return to antinumeracy, but first let's talk about symbols.

1.4   Symbols and the Logic Theorist

While Samuel's checkers program is often the first AI program cited, another candidate is the Logic Theorist written in 1956 by Allen Newell and Herbert Simon of CMU [75].

To explain the Logic Theorist we need to back up a bit. A *logic* is a formal system for reasoning from assumptions to conclusions. *Propositional logic* is one of the simplest logics, consisting of atomic propositions (typically named with letters **a**, **b**, etc.), one or more rules of inference, and a few *logical symbols* such as **if**, **and**, **or**, and **not**. With them we can form more complex propositions such as (**and a b**), which is true only if both **a** and **b** are true. Another example is (**if a b**), which says that if **a** is true, then it must be the case that **b** is true. If we were writing in English, of course, we would express the **and** statement with the "and" in the middle, as in "I went home *and* fed the dog," while the "if" would go before the two subsidiary statements. In logic, we simplify by always placing the connective before the two things being connected. As for the parentheses, think of them as serving the same purpose as a period at the end of a sentence: they separate one expression from the next. Also, we can compose such expressions ad infinitum: for example, (**and** (**not a**) **b**) says that **a** is not true, but **b** is. (This example also shows the use of the parentheses in that they clear up any possible misunderstanding that, for instance, it is not the case that the **not** applies to both **a** and **b**.)

Finally, there are the rules of inference. The classic rule of inference is *modus ponens*. Modus ponens states that if you know (**if a b**) is true, and you know **a** is true, then you may conclude that **b** is true. Another such rule is if you know **a** is true, then you may conclude **a or b** is true (since for **a or b** to be true, only one of its two arguments needs be true). The classic work on propositional logic is Russell and Whitehead's *Principia Mathematica*, which gives proofs of many theorems in propositional logic. Most works on propositional logic do not give many illustrations of proofs, but the whole point of *Principia* is to actually build up all of mathematics from first principles. So, if you are looking for a canonical set of propositional logic theorems, *Principia* is the place to go. However, don't ask me exactly how many are there. *Principia* is famous for being cited but never read [107].

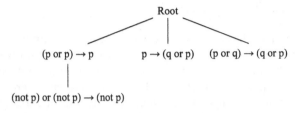

Figure 1.2: A tree of propositional logic theorems

The program that Newell and Simon built in 1956 was designed to prove propositional logic theorems from *Principia* and succeeded in proving thirty-two of them. It too characterized the problem as one of search. Just like checkers became a search through the game tree of moves, Newell and Simon realized that theorem proving could be implemented as a search through a tree of theorems, as shown in figure 1.2. A *theorem* after all is just the mathematical term for a logical conclusion. To create a tree of them, we start with all the axioms of the system. Perhaps one might be (or a (not a)), that is, any proposition is either true or false. (This is called the "law of the excluded middle.") Then the rules of inference allow us to construct new theorems, one or more for each such rule. If we know a, b, (if a c), and (if b d), then modus ponens allows us to infer both c and d. The point is that if we do this exhaustively we must eventually find the desired theorem in the tree. Newell and Simon realized they needed some heuristics to guide the search, which they had, but I am going to skip their exact nature.

I have spent time on the Logic Theorist, first for its historical importance, but also to reinforce the importance of symbols and symbolic reasoning in early AI research. Recall that a *symbol* is a "mark or character used as a conventional representation of an object, function, or process." The Logic Theorist and propositional logic are symbolic in the purest sense. Also Newell and Simon were the most explicit about the importance of symbols, having been the first to enunciate and embrace the "physical symbol-system hypothesis":

> Processing structures of symbols is sufficient, in principle, to produce artificial intelligence in a digital computer and that, moreover, human intelligence is the result of the same type of symbolic manipulations. (https://www.britannica.com /technology/artificial-intelligence/Methods-and-goals-in-AI)

1.5   Scene Recognition

Newell and Simon showed their bias toward symbol manipulation by their choice of research area. If you want to write a program to prove propositional logic theorems you must, by its very nature, write a program with emphasis on symbol manipulation. But even in areas where you might naturally emphasize numeric manipulation, the most interesting work was remarkably nonnumeric.

Consider the general area of scene recognition—going from light intensities as picked up by a TV camera to statements about the kind of objects in a scene and their locations. The most notable results in this area from this period were not related to the shape or size of objects— tasks that require numeric information—but the separation of straight-edge blocks in line drawings. In figure 1.3 (left and right), we see scenes that are quite similar, yet the first shows one block, while the second two. The task is labeling regions with distinct symbols, one for each separate physical block. The key idea is to look at the vertices created by intersecting lines. In these scenes a line is created in two situations. First, it may bound a block. The line A-L on the left separates the single L-shaped block from the background. On the right it also separates a block from the background, but the block is rectangular. The second line-creation possibility is when there is a crease in the block. In the left-hand figure, the line Y-B is a case in point. The regions on either side of this line belong to the same block.

The claim is that the shape of a vertex (and which vertices are connected) places constraints on how block surfaces group together into blocks. The shape of a vertex is the number of lines that emanate from it and the directions they head in. Some of the vertices in figure 1.3 are

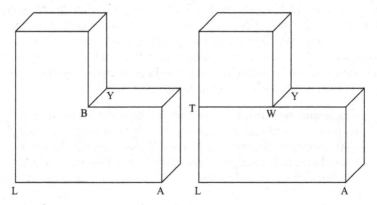

Figure 1.3: Similar scenes with one and two blocks, respectively

labeled with letters that express their shapes, such as A (for arrow), L, Y, T, and W. Note, for example, the arrow (A) vertices in figure 1.3. Several such vertices appear in the figure. Concentrate on the line that forms the "shaft" of the arrow, and note that in every case the regions on the two sides of the shaft are part of the same object. In fact, this need be the case, and using clues like this we can pretty much correctly identify the separate blocks in a scene. Thus, the input to such a block-separation program is a list of vertices, their labels, and their interconnections. No numeric information is required.

The first work in this vein was the SEE program [32] by Adolfo Guzman. A year later David Huffman in the United States [42] and Max Clowes in Great Britain [17] put this topic on a more mathematical footing. This was followed by David Waltz's 1972 MIT work extending the analysis to include shadows [106]. Waltz's work was also important in that it was one of the first applications of constraint propagation.

1.6 Constraint Propagation vs. Backtracking

Constraint propagation is worth a diversion because I am not going to get back to it later, and it is a technology that is widely used across CS. As such, it is a credit to AI for having such a large part in its creation, even if (at least in my estimation) it is no longer that important to AI itself.

In part I justified delving into checkers because it was a good illustration of search in AI. However search in checkers has its differences from search in other AI domains. For one thing, in checkers we are restricted to *sequential decision-making*. We choose a move (one decision), and it can only be after our opponent makes a move that we can decide on our second move. The player cannot make all the decisions at the same time, and the program cannot "change its mind."

Labeling each blocks-world area as either the background or part of an individual block requires a decision for each pair of surfaces—are these two surfaces part of the same block or not? At first, one would think that you should again make the decision sequentially since the first decision could well affect the second one (certain choices for the second one could be incompatible with the first). But unlike moves in two-player games, decisions in block-scene analysis can be retracted should other parts of the scene contradict the choice just made. This naturally leads to a style of search known as *backtracking*. If you hit a contradiction, go to the last choice (of region to block assignments, in this case), change it, and redo the subsequent analysis.

For block reconstruction, although it might be useful to make the decisions sequentially, it is not required, and, in principle, we could make many decisions in *parallel*—at the same time. This naturally leads to a process called *constraint propagation*. We mentioned earlier that the "shaft" of an arrow vertex must have surfaces of the same object on either side. This places "constraints" on the interpretation of the vertex at the other end of the shaft since there, too, the regions must be from a single block. So, initially we simply remove all vertex interpretations that are immediately inconsistent. This is much less computationally expensive than backtracking and, for the machines of that era, reduced processing times from minutes to seconds. (It is possible that the constraint propagation phase does not lead to a unique interpretation of the scene, in which case backtracking is used to help finish off the interpretation.) Constraint propagation is now well studied in the vibrant field of constraint satisfaction.

1.7 Natural-Language Processing and SHRDLU

In AI it is commonplace to distinguish human languages from programming languages by referring to the former as *natural* languages. So, for example, *machine translation*—enabling a computer to translate from one language to another—is placed under the rubric *natural-language processing* (NLP). Machine translation work started very early, in the 1950s, but was a noted failure from the beginning. See chapter 7 for more on why the results were so awful.

Another example of an early NLP system was that of Daniel Bobrow who in 1964 wrote the STUDENT program, which solved algebra word problems. However, the most striking success in early NLP was the SHRDLU program. (SHRDLU are the seventh through twelfth most frequent letters in English. The most common six are ETAION.) SHRDLU was the name Terry Winograd gave the program he developed at MIT in late 1960s. SHRDLU [110] took as input English statements about a simulated *blocks world*, shown in figure 1.4. This was "simulated" insofar as it existed only on a *cathode-ray tube*. (In those days, computer monitors showed only text, so specialized devices like cathode-ray tubes were developed for showing graphics.)

SHRDLU engaged in a conversation with the user. The prototypical example that appeared in Winograd's PhD thesis starts out like this:

1. Pick up the big red block.
   OK (see "gripper" pick up a block)

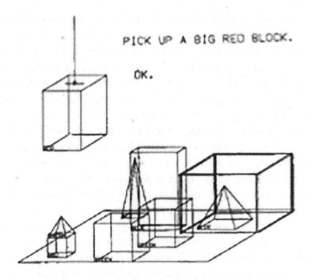

PICK UP A BIG RED BLOCK.

OK.

Figure 1.4: Blocks world for SHRDLU

2. Grasp the pyramid.
   I don't understand which pyramid you mean.

3. Find a block which is bigger than the one you are holding and put
   it in the box.
   By "it" I assume you mean the block that is bigger than the one
   I am holding.

4. What does the box contain?
   The blue pyramid and the blue block.

Here each numbered line shows a user input and SHRDLU's response.
The conversation goes on for another forty-three commands, queries, and
statements.

SHRDLU struck a chord in the research community for several reasons, not least of which was that Winograd's PhD thesis (which was
published as a complete single issue of a journal) was understandable
and well written. (And, of course, words are the prototypical symbols.)
More important, however, was that it was a successful program that
embodied a methodology for future research: (a) pick a limited domain
and simplify the comprehension problem by restricting the language to
sentences that directly apply to the domain; (b) leverage all aspects of
language, specifically including the meaning of sentences to achieve a
remarkable level of continuous dialogue between person and machine;

and (c) use a combination of logic and programs to represent the blocks-world knowledge required. We concentrate on the last of these.

SHRDLU adopted the standard separation in linguistics between *syntax*, *semantics*, and *pragmatics*. In linguistics, pragmatics studies how language interacts with the world. In SHRDLU the most striking interaction was the translation of English-language commands into logical instructions for the program to carry out. (A separate, less interesting program translated the movements of the "gripper" and blocks into cathode-ray tube instructions to make the movements visible.) The blocks world had its own "physics"—rules about how blocks can move. For example, no block can move if other blocks are on top of it. So, if you tell the system to move a block and it is not clear (free of blocks on top of it), you will see the offending blocks being moved elsewhere in the scene. To express the required knowledge and apply it when necessary, Winograd adopted *Planner*, a cross between a programming language and *first-order predicate calculus*. We will ignore the "first-order" bit (it is important, but peripheral to our discussion) and just call it the "predicate calculus." (By the way, this has nothing to do with "the calculus" that one might take in high school. "Calculus" comes from the Latin "to calculate" and can be used for things that enable any sort of calculation, including inference.)

The predicate calculus is a mathematical logic constructed on top of propositional logic. It was specifically designed for representing facts about mathematics but has also been pressed into service by AI researchers. It has the same logical connectives as propositional logic (**and**, **or**, etc.), and its rules of inference are a superset of those for the propositional logic. Where the two logics diverge is that propositional logic has *atomic propositions* (remember the **a**, **b**, and **c**s), while in predicate logic the propositions (or *statements*) have structure. The propositions are intended to assert facts about the world (perhaps just the mathematical world), such as:

(prime 17).

This would assert that the number 17 is prime. Here **prime** is a *predicate* in that it "predicates" something about 17, which is an *object* in the world. In this context, 17 is also an *argument* of the predicate **prime**. As in propositional logic, connectives and predicates go first, and parentheses are used to group individual statements to show how they build into combinations.

Lastly, predicate logic has *variables*. Using the notation common in AI, any string that starts with a question mark is a variable. For example, consider the following:

```
(if (and (even ?x) (not (= ?x 2))) (not (prime ?x))).
```

This states that anything that is an even number, other than 2, is not a prime. You should think of the if section (even ?x) as first "matching" a statement like (even 6). When it does so, ?x is *bound* to 6. Then, when we consider the second statement, the previously unbound variable now is replaced by 6 and becomes (not (= 6 2)). (If you are familiar with the predicate calculus, the variables here work as forall (x) expressions.)

Planner, as I mentioned earlier, is a component of the system that accepts queries in a cross between the predicate calculus and a programming language. So, if you asked SHRDLU about the color of the block in the box, it would construct a Planner program that looked like this:

```
(and (block ?x)
     (in ?x box1)
     (color ?x ?c))
```

This program would first look for all propositions in the database that match the expression (block ?x) (And, as we mentioned earlier, in so doing, bind ?x to the name of the block, e.g., block-1.) Then it would look to see if (in block-1 box1) is in the database. If it is not, Planner would backtrack to the first expression and see if there was another block statement in the database. It would continue to do this until it found a block for which the second line was true. Once it had, say, when ?x was bound to block-2, it would then check that the third expression could be found (for block-2) and get the value for ?c, its color.

I should mention that Planner was invented by Carl Hewett. Carl was another graduate student at MIT at the same time Terry Winograd and I were there. At this time Terry, Gerry Sussman (to be introduced shortly), and I were officemates, about five doors away from Carl, so we were all familiar with Planner, and all of us were interested in using it. Unfortunately Carl was good at thinking up great ideas, but not as great at writing programs, and he found elaborating Planner more interesting than coding it. So Terry, Gerry, and I built a quite primitive version of Planner, called "Micro Planner." (Terry and Gerry really—I just did some of the simple stuff.) Carl was miffed at first, thinking we had stolen his thunder, but the first line of the Micro Planner documentation cites him, and eventually he came around to the position that we had done him a favor by making Planner "real." To this day Micro Planner is the only version of the program extant.

Without a doubt, SHRDLU was the most successful piece of early AI in the area of language processing, but it did not provide a path for future progress. First, contained areas for language were hard to find. I wanted to emulate SHRDLU's success but in a slightly more complex "world." I tried textbooks for second graders. But the step between the blocks world and second graders was much too large. I got out with my PhD [14], but it was fair to say that my work was far from a success. (One kind, but blunt, senior researcher told me it was a "magnificent failure.") Nor did Winograd himself have much luck with the research program inspired by his thesis work. In 1985, with the philosopher Fernando Flores, he published a book [111] giving up on AI completely:

> We argue—contrary to widespread current belief—that one cannot construct machines that either exhibit or successfully model intelligent behavior.

Winograd then switched his attention to human-computer interaction, the branch of computer science concerned with making computers easier for people to use. He is a pioneer in that area.

I would like to give one last example about AI's early antinumeracy before moving on, this time a story against myself. As I briefly mentioned earlier, for my PhD thesis I was trying to write a computer program that could understand children's readers. Naturally one of my concerns was representing knowledge of the world a computer would need to reason about the situations children encounter in these stories. One day I was thinking about expressing knowledge about size, for example, that a particular ball was a large one. A standard predicate calculus representation might look like this:

`(size ball-2 large)`

At any rate, it occurred to me that one alternative might be to express sizes in, say, centimeters and express the information as

`(size ball-2 100)`

where 100 was, say, the diameter of the ball in centimeters. (To my physicist mind, using metric units is much more scientific than using imperial units such as inches.) However, even if the numbers were in centimeters, I felt resistant to using a number rather than a symbol. Somehow the symbol `large` seemed more informative than 100.

It took me a while to convince myself that I was being ridiculous. To a computer the string or symbol `large` does not have any particular meaning. If anything, it has less meaning than 100. A computer knows that $100 > 50$. It does not know that `large` > `medium`. Why not use the numbers and leverage this knowledge? It took me a few minutes to convince myself of my initial stupidity.

1.8 Learning in Early AI

AI today is almost coextensive with machine learning, and deep learning in particular. So far I have downplayed this area of AI, and to a large degree this is justified, because at the time we are talking about here, learning was indeed far down on the list of hot topics. However, there was more interest than I have indicated so far.

For example, after Samuel got his checkers-playing program working, he created several versions that incorporated various learning mechanisms. The simplest was to record the values found by minimax search. Suppose in the course of a game it is my move. I am at some position, call it position A. After I do the minimax search and static board evaluation (see section 4.1), I find that I can reach a board position with value +2. I will, of course, use this information to inform my choice of move, but I will also record that position A is worth +2 to me, so if I encounter it in some later search, I can use the +2 value and get, in effect, a deeper search. As we have already commented, the deeper you search, the better you play, and, indeed, when this feature is added, the checkers program plays better.

A second, more sophisticated form of learning is to modify the static board evaluation function. A simple evaluation function assigns a value to each piece on the board, and in our discussion we implicitly assumed they were all the same since to get a score from the perspective of black, we subtracted the number of red pieces still on the board from the number of black ones. This is equivalent to considering each piece on the board as having the value of 1. In a more complicated learning scheme, we make values of pieces different, perhaps based on position. Generally, pieces toward the center of the board are worth more, but by how much? One version of the checkers program used a simple form of learning to do the adjustments. Suppose two positions have the same value according to the current function, but one leads to a win and the other to a loss. To incorporate this information we could change the evaluation function so that now it evaluates the position that leads to the win more highly. This is a good idea, and we see it again in our discussion of AlphaGo in chapter 9. However, in Samuel's program it was difficult to implement, and the version that helped was complicated and *ad hoc*. Thus, while at the time we students were aware that the program existed in several versions, and the versions with learning did better, we were largely unaware of how those versions worked. I had to go to the Web to write these last two paragraphs, and if you think about them carefully, you might see holes in my description.

The only work on learning coming out of MIT in the '60s and '70s was Pat Winston's PhD thesis on concept learning [112]. His canonical

example was a program that learned the notion of "arch" when given
a high-level description of an arch, along with examples of configurations that were missing critical components, such as three blocks can
make an arch, but not if all three are on the ground. Winston's work
was well received, but in general learning was seen as a problem, not a
solution. My understanding of the MIT zeitgeist with respect to learning
was expressed as follows: "Who is smarter, you or a computer? Well, I
suppose I am. OK, now suppose you want to get the computer to diagnose a disease. Who is more likely to figure out how to do it, you or a
computer?" I am not sure who said this to me. My memory was that it
was Marvin Minsky, but it may well have been another student. At any
rate it was the common wisdom in those days because when put this way
the argument is quite compelling.

Indeed, I still think about where and how it goes wrong. I think (part
of) the correct response is, "Yes, but a person can try perhaps only a
half-dozen possibilities, while the machine can try millions or billions."

1.9   Perceptrons

The one major exception to this de-emphasis on learning was work on
perceptrons. Perceptrons are abstract neurons. As had been known since
the pioneering work by Santiago Ramón y Carjal, the brain is made
up of individual cells, like the rest of the body. (Prior to Carjal, most
anatomists thought the network in the brain was continuous.) There are
many types of *neurons*, as the cells have been named, but the prototypical neuron looks like those shown in figure 1.5. Neurons carry electrical
signals and typically have many inputs, called *dendrites*, and one output,
called an *axon*. The neurons carry the electrical signal to the *cell body*,
and if the signal is sufficiently large, the neuron fires, in which case an
electric signal propagates to the axon. The axon has many connections to
the dendrites of other neurons, but the connection points are *very* small
and were hard to see with the early twentieth-century microscopes, which
is how the cellular/continuous controversy arose.

In the early 1940s, Warren McCulloch and Walter Pitts invented
artificial neurons [66]—what are today are called "perceptrons." (This
is Frank Rosenblatt's name. We encounter him shortly.) Perceptrons are
typically drawn as shown in figure 1.6. As with neurons they have many
inputs and one output. Associated with each input the perceptron has
a number, called a *weight*. Each input itself is a number, and the two
numbers (the input number and the weight) are multiplied, and all the
products are then added together.

Perceptrons are *classifiers*. Suppose our inputs were images of single
digits (0, 1, 2 ... 9) and we want to design a perceptron to decide if the

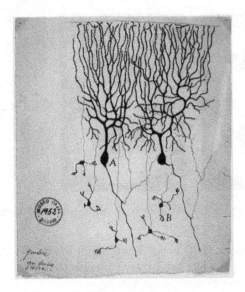

Figure 1.5: Two prototypical neurons

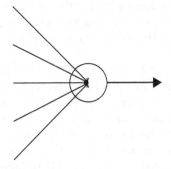

Figure 1.6: A single perceptron

input is an 8. This is a *binary classification problem* since the classification is between two possibilities: 8 or not an 8. We just described the computation done by a perceptron (multiply each input by its weight, etc.). By convention, if the resulting output number is greater than 0, the input is a member of the class.

So we are creating a perceptron to find images of 8s as shown on the left side of figure 1.7. The figure shows (center version) how this image would be translated into a version that a computer can interpret. If your eyesight is no worse than mine, you can see the inputs take values between 0 and 255, where 255 means that spot is reflecting light back at maximum intensity. The light intensity is low in most places since the

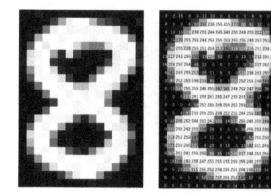 

Figure 1.7: A black-and-white image of an 8

background is black. This image has a resolution of $28 \times 28 = 784$ pixels, and a perceptron for processing such images would have one input for each pixel and one weight for each input.

So, to recognize 8s, we want the weights to be high in locations where an 8 has a line and negative where it does not. If the input is a 1, its lines would not overlap much with the lines of an 8. The 1's line's pixels would be multiplied by negative numbers (by the perceptron trying to recognize 8s), and thus, the 8 perceptron would correctly decide that the image is not an 8.

While McCulloch and Pitts invented the perceptron, it was not until the work of Frank Rosenblatt [84] that anyone actually considered using perceptrons for real-life classification. Rosenblatt devised the *perceptron learning algorithm*, which, given a collection of labeled examples, such as a set of images along with their *labels* (identification of the digit shown), could *train* the perceptron—set the weights to recognize the desired class. More precisely, if there exists a set of weights that enables the perceptron to associate the correct label with each image, the algorithm is guaranteed to find them. Unfortunately, for most real-world classes, there is no such set of weights. That is, for some sets the perceptron computes a number greater than 0 when the image was not an 8 and for others, vice versa. However, perceptrons are still often useful because, even when there is not perfect set, a good setting would get the vast majority correct.

The learning algorithm is remarkably simple. Start the perceptron with all 0 weights. Then, for each image, run the perceptron and get its prediction. Remember, a perceptron is a binary classifier with the rule that if the final value is greater than zero, the perceptron has decided that the image is a member of the class. When all the weights are zero, the output value is 0 no matter what the image, so the perceptron answers, "Not a member of the class." If, say, the image was a 4, then "not a member" is correct. When the perceptron is correct, we do nothing and go on to the next image/label pair. Suppose the second training example is an 8. This time the 0 answer signals a wrong classification. In the case of an incorrect answer, we go through each pixel value times weight value and modify the weight so that the perceptron's final value is closer to correct. In the case at hand, the image was the 8 in figure 1.7. If for each illuminated input we add a small amount (say, .2) to its weight, the next time we see this image, the new weights compute a value greater than 0 and thus classify this 8 correctly. If the mistake is in the opposite direction, then we subtract from the weight value.

The major limitation of perceptrons is that in most cases no set of parameters can capture the class we want to represent. Indeed, Marvin Minsky and Seymour Papert wrote a book, *Perceptrons*, that developed the mathematics of perceptrons and showed that its limitations are severe [73]. The book came out in 1967, the year before I arrived at MIT. It had the effect of shutting down perceptron research not just at MIT but pretty much everywhere.

Thus, from my vantage point at MIT, perceptrons were not a research option to be pursued. Also note that while the numbers would be set correctly, it was not obvious exactly *what* was learned. A story I have lodged firmly in my memory as being told by Minsky was that a perceptron was trained to distinguish images of Russian tanks from US tanks. Unfortunately, the images were from two different data sets, taken under different circumstances. The Russian images had lower exposures and, thus, were, in general, darker on average than the US tanks. In fact, all the perceptron ended up doing was computing the average light intensity and responding "Russian" if the value was low. (However, a website named The Neural Net Tank Urban Legend says so many different versions of the story exist, promulgated by so many different people, and thus the story is most likely not true. However, one of the individuals cited is Minsky, so it is plausible that I heard this from his lips. While not true, this story points to the real difficulty of evaluation in general and evaluation of neural networks in particular. A perceptron is a neural network, or NN, with just one "neuron." In chapter 7, I mention mistakes I have made in this vein.)

Before we leave perceptrons, let's look at one more diagram to prepare us for chapter 8 and *deep learning* where they play a starring role. Figure 1.8 shows a diagram of ten perceptrons (the ten boxes making up the thin rectangle on the right), each being fed with all the pixel values in the schematic image represented on the left. Note that all ten perceptrons are separate. They each have their own weights, and if you

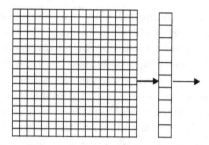

Figure 1.8: Perceptrons for recognizing all ten digits

like, you can think of them as being trained separately. The idea is that, say, the topmost is being trained to recognize 0s, the second from the top 1s, then 2s, and so forth. Then, whichever value comes out highest, that digit is the system's guess.

The ten perceptrons can be trained in parallel since, as I said, you can think of them all as separate. However, as is now standard, I have shown them as a single narrow box, a *layer* to use the terminology that is going to be standard fifty years hence when all of this becomes part of deep learning. However, in fifty years, there will be many layers—that is where the "deep" comes in. See chapter 8.

2

Reasoning and Knowledge Representation (1970–1985)

2.1 Reasoning, Problem Solving, and Planning

Reasoning is the process of reaching new conclusions based upon facts already in one's possession. Together with perception (visual, auditory, etc.), it is a fundamental component of intelligence. For this reason, as AI was settling down to become a standard branch of computer science, most of the canonical work of the time (1970s and early 1980s) fell under "reasoning," along with its sister topic, "knowledge representation."

In fact, another very early AI program that I have not mentioned yet was the General Problem Solver (GPS) by Allen Newell and Herbert Simon, of Carnegie Mellon and Dartmouth Conference fame. Problem solving in AI comes up in two guises—domain dependent and domain independent. In the former the program has been customized for a particular problem. We return to such programs when they appear under the name "expert systems."

We can also distinguish between *problem solving* and *planning*. In the former a program takes a description of a problem and solves it, whereas in the latter the goal is to produce a *plan* that, if executed, would solve the problem. The difference between these settings can be small, however. Consider a program to solve the sliding-squares puzzle. The version shown in figure 2.1 contains eight numbered squares on a 3 × 3 board. Initially the numbers are jumbled, and the goal is to get them in order by sliding the correct squares into the empty position. Figure 2.1 shows a typical starting state and the goal state.

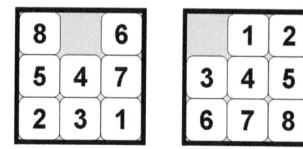

Figure 2.1: A typical starting state and the goal state for a sliding-square puzzle

In both planning and problem-solving research, nobody builds a robot to play this game—that would be considered robotics since building a robot hand to play sliding squares is *much* harder than solving the puzzle. Instead a problem-solving program would drive a simulation. But, of course, you can turn this into a planning program by tracking the moves sent to the game simulation and then outputting the list of moves all at once at the end as the plan. However, as we will soon see, the difference between the two problem settings can be important, and the field has standardized on planning as the problem of choice.

2.2   STRIPS

The founding work on domain-independent planning is the STRIPS system, the Stanford Research Institute Problem Solver, which, as you would expect, came out of the Stanford Research Institute (SRI). STRIPS was developed in 1971 by Richard Fikes and Nils Nilsson [28]. Also, illustrating our point about the small distinction between problem solving and planning, despite its name—Stanford Research Institute *Problem Solver*—STRIPS is a planner.

STRIPS was the "brains" of the SRI robot, "Shakey," so named because of its tendency to shake when in motion (figure 2.2). STRIPS was in charge of translating between high-level goals and more basic robot motion commands. For example, if Shakey were told "go to the red block" and that block was on a raised platform, STRIPS would figure out that the robot first needed to go up a ramp.

STRIPS is domain independent in the sense that a *STRIPS instance*— a problem to be solved—contains the necessary domain information, expressed in the STRIPS language, as well as a description of the current state of the world and the goal state. Figure 2.3 shows the start

Figure 2.2: Shakey the robot

and end states for a blocks-world problem. (Shakey did not have an arm, so it could not place a block, but, as we said, STRIPS is intended to be domain independent.) Figure 2.4 shows the STRIPS instance for this blocks-world problem. The action Move has as its precondition that block ?b is on something (?y) and is Clear (nothing is on it). There are also general sanity conditions, such as ?b and ?y cannot be the same block (not (= ?b ?y)). Note that I have used the traditional negation symbol, so this is expressed as ¬(b=y). STRIPS uses what is called the *closed-world assumption* in which a negation (e.g., the *not equal* precondition) is assumed true unless there is an explicit statement to the contrary. It also assumes that entities with different names are distinct (not equal).

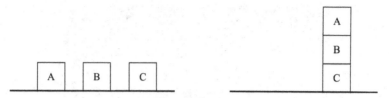

Figure 2.3: Start and end states for a blocks-world problem

```
Init    (On A,Table), (On B,Table), (On C,Table),
        (Block A), (Block B), (Block C), (Clear A) (Clear B), (Clear C)
Goal    (On A,B), (On B,C)
Action (Move ?b, ?x, ?y):
        PRECOND(On ?b,?x), (Clear ?b), (Clear ?y), (Block ?b), (Block ?y),
               ¬(b=x), ¬(b=y), ¬(x=y)
        EFFECT  (On ?b,?y), (Clear ?x), ¬(On ?b,?x), ¬(Clear ?y))
```

Figure 2.4: STRIPS formulation of the blocks-word problem shown in figure 2.3

As we saw with other inference problems, planning in STRIPS is reduced to search though a space of world configurations. In the simplest version, we start with the state of the world as provided by the problem instance and consider all possible actions whose preconditions *unify* (e.g., match) statements in the problem description. The process of matching is called *unification*, and a given action can potentially unify with the world description in more than one way, for example, Move in figure 2.4 can unify with the variable ?b matching, say, block A or B, and similarly with the variables ?x and ?y. We then create a search tree with an initial state whose descendants are states of the world after different actions, or different matchings of the same action. We create these states by considering all of an action's *effects* in turn, removing from the world state any statements that have become false (for example, a block might no longer be clear), and adding the statements that are now true. When we find a state in which all of the statements in the Goal section of the problem instance (and none of the negated ones) are present, we are done.

2.3   Ordering Actions

We have said that early AI did not fully appreciate the difference between planning and problem solving because it seemed that one could turn a problem-solving program into a planner by simply recording the actions

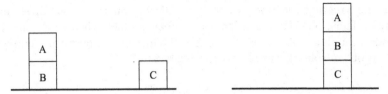

Figure 2.5: Start and end states for a creative-destruction problem

{ (On B C) (On A B) }

{ (Clear B) → (Stack B C) }
{ (Clear A) → (Stack A B) }

{ (Clear B) → {(Clear A) → (Stack A B)} → (Stack B C) }

Figure 2.6: Operation of NOAH on a creative-destruction planning problem, showing plans at three levels of elaboration

taken and returning the list at the end. This view presents some difficulties, however. A classic case arises with the problem illustrated in figure 2.5, which for a time was considered important enough to deserve its own name, the "creative-destruction" problem. The problem has two goals—having A on B and B on C—but superficially it appears that one of these goals has already been satisfied for you, namely, A starts out on top of B in the initial state. Actually, however, it is necessary to "destroy" this initial goal because the second goal, (On B C), requires undoing it since block B needs to be clear to be able to move it. Furthermore, since the planning programs of the time were not anticipating this, they would stop after the first action, thinking they were done.

It is now recognized that it is possible to solve this problem in many ways, but the first planning program to do so correctly was the NOAH program, created in 1975 by Earl Sacerdoti, who, like Fikes, was at SRI [90]. Figure 2.6 shows NOAH's operation. The basic idea is that NOAH has two goals initially, but they are unordered with respect to each other. In step 2, they are both expanded into their parts. Here, only the parts within a single action are ordered. Finally, because of the constraint on when you can stack a block, the (Clear B) from the top goal is ordered before all elements of the lower goal, which in turn comes before (Stack A B).

Like STRIPS, NOAH is performing a search to find a plan, but where STRIPS searches though a space of states of the world, NOAH is searching through a space of plans. Each of the three parts of figure 2.6 is a plan for creating the tower of blocks, differing on their level of elaboration. Thus, NOAH is definitely a *planner* in that, when it is done, it produces a plan. Subsequent to NOAH, the distinction between planning and problem solving was clearly established.

2.4   Planning Heuristics

One clever aspect of STRIPS planning is the interplay between the representation of the domain and search heuristics. Consider again the sliding-squares puzzle. This puzzle is often used as a test example for search and there are several good heuristics to address it at different levels of granularity. One, the misplaced square heuristic, simply counts how many squares are in their final position. A more accurate but slightly more computationally expensive measurement is the Manhattan distance heuristic. "Manhattan distance" is the effective distance required when one is allowed to move only directly north, south, east, or west, as in the borough of Manhattan. The measurement of how far each square is from its final place according to Manhattan distance is a lower bound on the total number of moves we must make to reach the goal state.

Now consider the following pointless "heuristic" for judging the quality of an intermediate state in a sliding-squares puzzle: solve the puzzle from that state, count how many moves it takes, and use that number. This method is pointless, because finding the heuristic's value takes as long as solving the puzzle in the first place. But suppose we do a simplified version of the search, one that ignores some of the action preconditions. This approach will make the search much shorter and thus provide information at a cost less than solving the original problem. In the case of sliding squares, the problem has one action, which looks like this:

```
Action (Move ?s ?p, ?q)
      PRECOND (At ?s ?p), (Adjacent ?p ?q), (Empty ?q)
      EFFECT  ¬(At ?s ?p), (At ?s ?q), ¬(Empty ?q), (Empty ?p)
```

(To move ?x from location ?p to ?q, the former must be adjacent to the latter, and ?q must be empty.)

Suppose we temporarily remove the requirements that the square we move to is adjacent to the square we are moving from—as if we pryout the square to be moved and place it wherever we want. In this "game," the number of moves required for a solution is the number of misplaced squares, since we would be allowed to pick up each square and put it in its goal position. In figure 2.1, moving the "3" to its goal position would take just one move. The same is true for all the tiles except "4," so the heuristic, and the number of moves in this warped version of the game, is seven. Judging the quality of a position now takes much less time than solving the problem. Or to put it slightly differently, relaxing the two preconditions in the search is almost equivalent to using the "misplaced square heuristic."

Alternatively, we could just cancel out the Empty precondition. In this version, the number of moves required is that returned by the Manhattan distance heuristic since here we pry the square out but must replace it in an adjacent position. So, for instance to move the 3 to its goal position would require two moves: one up and one to the left.

The correlation between search in the relaxed-precondition games and useful heuristics has led to research in the automatic discovery of search heuristics.

2.5   Expert Systems

An *expert system* is an AI program to aid making a decision normally made by a human expert in some area, so these systems too fall under the "reasoning" rubric. The heyday of expert system research ranged between 1970 or so to the mid-1980s. Arguably the first expert system is the Heuristic DENDRAL program, created to find the structure of a chemical molecule. Given a molecule's composition, such as $C_8H_{16}O$, it would determine which atoms are connected to which, as in:

$$CH_3 — CH_2 — \overset{\overset{\textstyle O}{\|}}{C} — CH_2 — CH_2 — CH_2 — CH_2 — CH_3.$$

The idea of the program was primarily that of Stanford's Joshua Lederberg, who won a Nobel Prize for his work on the genetics of microorganisms. Later in his career he became interested in the use of computers for scientific discovery. For this project he recruited both chemists and AI researchers, with the latter being Bruce Buchanan and Edward Feigenbaum, also of Stanford. Their work came out in 1969 [11].

Besides the formula, which just lists the names and quantity of the atoms present, the program is also given the output of a mass spectrograph of the molecule, as shown in figure 2.7. A mass spectrograph takes a sample of the molecule in the form of a vapor and bombards it with subatomic particles, which causes the molecule to break into pieces. The pieces are usually positively charged, having lost one or more electrons in the process. By sending the stream of molecular pieces through

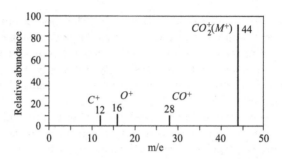

Figure 2.7: Mass spectrogram for a simple molecule

a magnetic field, the scientists can measure the ratio of fragment mass to fragment charge and this is what is illustrated in figure 2.7. Since the charges are always small (typically one or two units) and since the places where a molecule will break are reasonably well understood, the molecular composition together with the mass spectrograph are often sufficient to determine the structure.

Perhaps by virtue of being an early example, Heuristic DENDRAL is not a prototypical expert system. Its chemical knowledge is largely built into the structure of the program, and for this reason, I will not dwell on how the program works. A better example for our purposes is the R1 program, which was built and used by the Digital Equipment Corporation, (or DEC for short) in about 1980 [68]. The program was written by John McDermott of CMU who reportedly gave the program its name because "Three years ago I couldn't spell knowledge engineer, now I R1."

R1 configured DEC VAX computers. In the days before personal computers, so-called *mainframe* computers had to be *configured*. Concretely, ordering a mainframe computer typically involved ninety items, each purchased separately, and it was common for orders to not include all the required pieces. R1 (renamed XCON) was one of the few expert systems that was, in fact, routinely used to solve the problem for which it was created.

A canonical expert system is created by the programmer in conjunction with domain experts coming up with many *if-then rules*. R1 had about 2,500 of them. If-then rules have the form "**if** *such-and-such* is true, **then** *do something*,"so an actual rule might look like this:

```
if (and    (contain ?order printer-type-1)
           (not (contain ?order cable-type-3)))
then add
           (contain ?order printer-cable-type-3).
```

I made up this rule, but actual R1 rules were more complicated. My rule is designed to see if an order contains a particular type of printer but does not contain the corresponding cable. If the condition matches, the rule says to add the cable to the order. Note that the rule does not specify the order number. Rather it contains a *variable*, ?order, which, as in STRIPS, matches anything. (Adding a question mark before a symbol was a typical way to indicate that something is a variable.) An expert system has a database of such rules and a (possibly separate) database of facts about the order. We say that a rule is *applicable* if its "if" part (called its *left-hand side*) is true. In this case it requires that the database contain the proposition

```
(contain order-5 printer-type-1).
```

When we match the first part of the if statement, its variables are bound to symbols in the database proposition—in this case, ?order is given the value order-5. Also, again like STRIPS, not statements are considered true if no database statement matches the negated proposition. The program repeatedly looks for rules whose left-hand sides are true and then adds the information on the right-hand side of the rule.

Programs with this sort of architecture are called *production systems*, and the individual rules are called *production rules*. They support *pattern matching* and the retrieval of propositions that could match a particular propositional pattern, such as finding the statement about order-5 from the version with the variable.

It is not hard to see why the production system architecture proved so popular. In the ideal case the creators don't have to be programmers at all. Rather, domain experts write the domain rules, and they are done. Unfortunately, this ideal is never achieved (or, at least, I have never heard of a case when it was). Given an extant production-system architecture, it always seems that a programmer must add a new feature or work around issues by what can be called "misusing" the system. For example, if our rules are just dealing with facts and inferences from some facts to new facts, it should not matter in what order we make the inferences, except for the obvious requirement that if A is inferred from B, then B must be established before A. But R1 rules, in fact, have gating clauses. The set of conditioning facts for the application of a rule typically includes a clause like (working-on printer-connections). And another rule might, in effect, say that once you have finished with the printer connections, start working on printer internals. Typical computer programs, of course, require the ordering of most everything, so these gating conditions mean the domain expert must be somewhat of a programmer as well.

R1 was not an early expert system. I chose to consider it because its problem domain (VAX computer configuration) allows us to ignore a complication that I want to consider now. The next important expert system was the MYCIN program, developed at Stanford by Edward Shortliffe and Bruch Buchanan in 1975 [12].

MYCIN's area of expertise is bacterial infections. It queries a physician, and based upon the answers, it identifies the bacteria observed in the patient and recommends a course of treatment. In our discussion, we will consider only the identification phase.

MYCIN also uses a production system, with one major difference from classical production systems like R1. The information that doctors, and, therefore, MYCIN, have at their disposal is often not sufficient to uniquely determine the bacteria causing an infection to the exclusion of all others. Therefore, MYCIN's rules do not reach necessarily true conclusions. Instead, the rules must accept uncertain premises and express how certain one may be in its conclusions. This proved difficult—so difficult that it is going to need an entire chapter, the next one, to resolve.

2.6   Knowledge Representation

Both planning programs and expert systems use propositions to represent what they know of the world. Classical planning is indeed partially defined by this use. (We'll see another form of planning called *reinforcement learning* in section 9.3 that does not necessarily use propositions.) Similarly, any expert systems that uses production systems must also use propositions.

The use of propositional knowledge entails the use of predicates, such as On, or Clear, and unique objects like table. The study of which ones are needed is called *ontology*, a term imported from philosophy, where it denotes the branch of metaphysics dealing with existence, or the nature of reality. The term "knowledge representation" in AI is slightly broader, including ontology, but also enough logic to support inference, the idea being that it is the inferences in which a term participates that pin down its meaning.

In the early days of AI, say, the 1960s and early '70s, a fair bit of confusion existed about what counted as different representations, most notably the supposed distinction between propositions and "semantic networks" or "semantic nets." A semantic net is a graph in which the nodes are concepts, or objects, and the edges are the relations between them. A prototypical example might look something like figure 2.8, which is intended to mean that Tweety is an object of type bird, and birds can

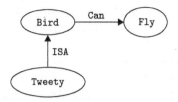

Figure 2.8: Typical semantic network

fly. But when thought about, the question arose, well, exactly how is this different from

```
(ISA Tweety Bird)
(Can Bird Fly).
```

Eventually, researchers realized that the only real difference was that the semantic net version suggested some sort of indexing, whereby it would be possible to find what kind of object Tweety was once you had the symbol `Tweety`. However, remember that when we discussed production systems, we said they supported retrieval of propositions from their partial descriptions, as in (ISA Tweety ?x). The standard way this would be accomplished would be to have associated with `Tweety` all of the propositions that use it. Thus, it was realized that the two notations were equivalent, and everyone converged on the propositional version.

A similar situation arose with so-called *frames* or *scripts*. Both terms applied to groupings of facts larger than individual propositions. Roger Schank along with Robert Abelson, both of Yale, advocated the use of *scripts* in AI to refer to a statement of how an action is to be performed [92]. Their canonical example was eating at a restaurant—the *restaurant script*. In retrospect, this was not that different from the planning information used by STRIPS, but the purpose was different. Schank and Abelson were more concerned with plan recognition than planning—that is, if you read that Sally and April ordered steak for lunch, then you know they are participating in the restaurant script.

Marvin Minsky was responsible for *frames* [72], which were in many respects similar to scripts, though a frame could be a collection of knowledge about an object (for example a piggy bank) as well as actions. Again, eventually researchers decided that these terms denoted organizational issues and could simply be thought of as collections of logical statements.

We noted at the outset of this section that knowledge representation encompasses both ontology and a logic to support inference from the expressed knowledge. STRIPS is a logic since it is a formal method of making inferences, so if your knowledge was in service of a STRIPS planner, your logic was already fixed. However, STRIPS is quite limited as a logic. Instead, since the predicate calculus already existed, it was usually assumed as the logic of choice.

I introduced predicate calculus in my discussion of the SHRDLU language-understanding program, but really, the logic used therein is almost as limited as STRIPS. Once you start taking the predicate calculus seriously, you realize that it is nearly impossible to use it to express the kinds of inferences used in everyday reasoning.

For example, consider the facts we included in the small semantic net of figure 2.8. Before we can conclude that Tweety can fly, we need a rule that says

```
(if (and (ISA ? x ?type) (Can ?type ?action))
    (Can ?x ?action))
```

Here, the **if** clause will access the two propositions expressed by the semantic net, and the **then** portion of the rule allows us to conclude that Tweety can fly.

But suppose something is wrong with one of Tweety's wings and we know that

```
(not (Can Tweety Fly)).
```

In the predicate calculus, this is not allowed. Remember the law of the excluded middle. In the predicate calculus, a proposition must be either true or false, but it cannot be both. In this case, we can infer that Tweety can fly (because she is a bird), and are directly told that she cannot. A possible interpretation of this situation is that prior to being told that Tweety cannot fly, the logic supported the inference that she could fly, but this changed when we were explicitly told otherwise. Logics that have this property are called *nonmonotonic* logics. Calling something "monotonic" means it is always heading in the same direction. Predicate logic is monotonic in that the set of inferred statements never decreases— by adding new statements you may increase the set of statements that may be inferred, but no inference goes away. Several nonmonotonic logics were invented in 1980 [63, 67, 82].

A second major subeffort within knowledge representation is the *axiomatization* of commonsense knowledge. To "axiomatize" a body of knowledge is to write down the facts needed to infer important conclusions in that area, and by "commonsense" knowledge, we mean facts about the everyday world: how to order food at a restaurant, not how to build a suspension bridge. (An "axiom" is the term in logic for a basic fact assumed to be true.)

One example was suggested by Pat Hayes at the University of Rochester in "The Naive Physics Manifesto" [33]. Here he put forward a research program for formalizing commonsense physics, as opposed to "real," mathematical physics. The project aimed to simplify the problem by not asking the researcher to produce a working program that, given their formalization, could derive the desired conclusions, but rather simply checked "by hand" that they would. Hayes also produced an example of what he had in mind by formalizing commonsense knowledge of liquids [34]—a hard area because of the way puddles can form, split, evaporate, overflow, and so on.

This research program struck me as quite reasonable. The axioms that Hayes came up with were complicated but plausible. Further, it was obvious what to do next: repeat the exercise for different domains. A good friend of mine, Drew McDermott (at Yale), had his first graduate student, Ernie Davis, work on how solids would move with respect to other solids [22]. (A small pebble in a large funnel was his example.) However Davis's work was, for me, somewhat sobering. First, it was hard. I eventually tried to do something similar but with painting things. I could only write rules that would allow me to come to the most trivial conclusions. Furthermore, I could not see how any of the Hayes or Davis work could be a foundation for mine. Rather I felt I was starting from scratch.

2.7 Reasoning and the Symbol System Hypothesis

Classical planning by definition adopts the physical symbol system hypothesis by representing the state of the world in terms of predication logic, as well as goals, actions, and everything else. The same can be said for expert systems, at least those that adopt the production system architecture. However, those of you who read the preface of this book know that this author sees the future of AI in the new deep learning camp, which tries to minimize or eliminate the use of symbols. That's because, at least so far, they do not work well with deep learning methods. Thus, the question arises: Does the research covered in this chapter argue against this position?

To a surprising degree, the answer is "no." Most notably the topic of expert systems has faded from view. It is not that AI researchers have stopped working on programs to solve difficult problems in the sciences, but rather that these programs no longer use production systems. As we see in chapters 10 and 11, there have been recent success in AI solutions to interesting problems in chemistry, mathematics, physics, and, most notably, biology, where the program AlphaFold has solved a forty-year-old open problem: infer the shape of a protein from its building blocks.

But as all of these programs use deep learning methods, they support, not contradict, the thesis of this book.

Nor has knowledge representation been notably successful. In the writing of this book, I have been using the textbook *Artificial Intelligence, a Modern Approach* by Russell and Norvig [88] as my informal guide to what constitutes "mainstream" AI. Their position is this:

> We should say up front that the enterprise of general ontological engineering has so far had only limited success. None of the top AI applications make use of a general ontology—they all use special-purpose knowledge engineering and machine learning [88, 318].

But what about planning? Planning has not been an area of my research, so when I started this history, I wondered if it is still an active research paradigm within AI. Perhaps it had gone away, like expert systems, in which case my story would be much simpler and much more convincing.

But this is not the case. First, like other areas of AI, planning has been professionalized over the years. For example, in 1998 the Planning Domain Definition Language (PDDL) was introduced as a common language for classical planning research. It is based on STRIPS and is used to define domains so that different groups of planning researchers can compare results. Since then the field has been informed by the results of the International Planning Competition, which uses PDDL. (Both PDDL and the competition began through the work of a committee headed by Drew McDermott.)

Furthermore, classical planning is used in the outside world. The National Aeronautics and Space Administration (NASA) schedules the actions for its missions (e.g., the Mars rovers) with STRIPS-like planners, as does the US Army for scheduling transport. I find the creation of heuristics from the relaxation of propositionally expressed preconditions and postconditions quite natural and elegant. So, planning is alive and well and until something is found to, in effect, replace it, or alternatively, find a way to merge the symbols it requires with deep learning, AI, in my estimation, does not yet have a satisfactory foundation.

3

Reasoning under Uncertainty (1980–1990)

3.1 Expert Systems and Uncertainty

A large fraction of expert systems are attempts to solve *diagnosis* problems—situations in which we have observed the world and attempt to reach conclusions about why the world is as it is. MYCIN's identification of the bacteria from test results is an example. In diagnosis, we reason "backward" from effects (what we observed) to causes, where the causes are not directly observable. The classic case is medical diagnosis. A patient comes to a doctor with various symptoms, and the doctor wants to diagnose the underlying cause, say, diabetes (a disease where the body produces too little insulin), since physicians are vastly more likely to ameliorate the patient's problem once they know the cause (for example, they can recommend particular medications or, more directly, inject extra insulin). The problem for the doctor, or an expert system, is that typically symptoms are not neatly paired with diseases. Many symptoms are associated with diabetes, and a patient typically shows some, but not all, of them. Also, not only can patients with different symptoms have the same disease but people with different diseases can have many of the same symptoms. Excess thirst is a common symptom of diabetes, but other issues could be causing the problem. So, we cannot reason

if we see the symptom frequent thirst, conclude diabetes.

Rather we need a rule like this:

see "frequent thirst" → consider "diabetes."

However, this raises that question of what we mean by "consider."

We want the answer to the question, "What is the disease causing the symptoms?," where the nature of the symptoms makes some answers more likely than others. Numbers are a good way to rank things, so associated with each possible answer is a number—a measure of belief. Historically, this measure is called a *probability*. Unfortunately, people find probabilities confusing, and perhaps for this reason, this chapter may be the most difficult in this book.

The mathematics of probability and its cousin, statistics, were invented to better understand games of chance. In a simple case, you draw a card because you have bet on the outcome. If the deck has fifty-two cards and we have shuffled the deck, there is one chance in fifty-two of drawing, say, a two of hearts. We speak of the card drawing as an *event*, an action with a fixed set of possible outcomes. Rolling a six-sided die is an event with six possible results. Associated with an event is the probability of each of the outcomes. The set of these probabilities is a *probability distribution*—a set of nonnegative numbers that sum to one— which is to say you cannot have a probability less than zero, and since the event must have *some* outcome, the total probability must sum to one. (I am not sure why, but for the longest time, I had a problem with this statement about probability distributions—I thought it was saying we could assign any set of numbers we pleased to an event, for example, (.05, .95) to the outcome of flipping a fair coin. Needless to say, that is not what we mean. Rather, it is saying what sets of numbers are *not* allowed, just like numbers for the brightness of a light bulb cannot be negative.)

Roughy speaking, statistics is the mathematics of estimating probability distributions, and probability is how distributions are combined when we have several events. Both theories are well articulated and deep, but when the issue of reasoning about uncertain events came up among AI practitioners in the 1970s and '80s, researchers had trouble applying the math to the problems at hand. The result was they rejected mathematical probability and statistics, instead building new systems that seemed to better meet their needs.

The first problem was that few of us had had good courses in probability and statistics; remember the antinumeric tendencies AI has had since its early days. Thus, early AI practitioners, when looking for mathematical backup, almost exclusively looked to mathematical logic. Certainly that is what I did in graduate school—it never occurred to me to take probability and statistics. Also, the folks teaching those courses never made much attempt to attract AI researchers back then. But physicists need mathematical statistics. With my physics background, why did I not have all the probability and statistics I needed?

3.2 Frequentist vs. Bayesian Statistics

Another reason AI researchers shied away from probability and statistics has to do with the problem that statistics has not one, but two competing underpinnings, called *frequentist* and *Bayesian*. We are talking now about the 1960s, and at that time, almost all (perhaps *all*) college-level courses were frequentist in orientation. As such, the material they taught was of marginal use to us AI folks but very useful to physicists.

Remember my formative experience in measuring properties of neutrinos at Argonne Laboratory. My programming problem was to cull from all the particle-particle interactions those that at least looked liked they involved the neutrinos we cared about.

The persistent problem is that random permutations of the experiment create situations that look like neutrino events but are not. Frequentist statistics helps the experimenter answer the question, "What is the probability that what we observed was, in fact, just due to such a random combination of events." If that probability is sufficiently small, we assume we are justified in believing the results of our experiment. Good experimental design requires the researchers to specify in advance what level of certainty they (and, in their opinion, the whole scientific community) require before the results are taken as "real"—not due to chance. For example, a decade ago, when particle-physics researchers were looking for the Higgs boson, a particle that should exist if the so-called *standard theory* of physics is correct, the experimenters specified that they wanted their data to have only one chance in a million that it could have been due to coincidence. This has been traditionally the standard for believability for a major result.

To be more specific, in frequentist statistics, we assume we have two theories, let's call them the old vs. the new. In this case, the old theory does not include the Higgs boson, but it does say with some probability we can still see events that look like a Higgs particle. We also have the Higgs theory. It predicts that true Higgs events occur. Since the false non-Higgs events still exist, the total number of Higgish events are larger than if the old theory is true. So, the question then is how many of these extra events do we need to see before it is very unlikely that we are seeing only false ones? When the probably that we are seeing just false events is less than one in a million, the researchers and their colleagues switch their allegiance to the new (Higgs) theory.

Obviously, this does not sound at all like the diagnosis problem faced by an expert system with a medical diagnosis problem. It is not as if the patient will go into the doctor's office over and over, giving us the multiple versions of signs and symptoms that we need for our statistics. There is, however, a second way of thinking about statistics, called *Bayesian statistics.* Bayesian statistics does not assume that we can get probability distributions only though repeatable experiments. Rather, we allow distributions to be *subjective.* Suppose you are an experienced physician practicing internal medicine. As such, you have seen a lot of patients with diabetes, and, if asked to make a guess, you might say that one out of every 500 patients suffers from this problem. To take an example closer to my experience as a college professor, if a student makes an appointment to see me (and I do not, right now, know anything more specific about the student), I would guess with probability one out of five that they are seeing me because they are having trouble with a course I am teaching. I would not make the probability much higher—they may be coming in because they would like a letter of recommendation for graduate school, or they are graduating and stopped by to say farewell. So, if we let A be the appointment event, $P(A = \text{problem}) = .2$ is a subjective estimate that means the probability that the appointment is about a problem with the class is 20 percent. It is not exact, but it is also far from random. However, it is still the case that this is a number I made up.

Because Bayesian statistics deals with "made-up" numbers, until recently, hardly any statistics departments taught the subject from a Bayesian point of view. I remember a conversation with a more theoretically oriented AI researcher who said that frequentist statisticians thought that Bayesians were little better than witch doctors. (This was back about 2005.) As he uses statistics for AI problems, he is a Bayesian, and this was said with more than a bit of animus.

3.3 Probability

The resolution to the issue of handling uncertainty in expert systems and the rest of AI lies within standard probability, so before we go there, first a brief introduction to the mathematics involved. It turns out that the math we need is pretty much independent of the statistical perspective, Bayesian vs. frequentist. I find it helpful to keep in mind, however, that the probability distributions we are talking about are not objective in the sense that we measured them but are simply beliefs of the designer of the program.

Before, we spoke of "events" as actions—picking a card from a deck, colliding two particles. We also allow less "action-like" cases to count as events. Tomorrow's weather can be an event, with the outcomes, say, fair, cloudy, rain, or snow. Thus, we speak of a 50 percent chance of rain tomorrow. To apply probability to language processing, such as to machine translation, we consider the translation or generation of a sentence to be a sequence of events—generating the first word, generating the second word, and so forth. We might guess that the first word in the next sentence I write is "The." That is not a bad guess since "The" is, indeed, the most common first word of a sentence, but we would have lost the bet in this case. If we take an English word to be any string of characters that appears in an English sentence between two white spaces, there are arguably an infinite number of words. (We see new proper names all the time.) We may have an event with infinitely many different outcomes, but things are simpler if they are finite, so often we fix our English vocabulary at, say, 50,000 of the most common words. In chapter 10, we discuss a better solution to this problem.

To write about events more clearly, we give them names, typically single capital letter, like "E," and to specify the outcome, we say, for example, $E = 6$ if the die came up 6. Or, if "F" is the name we give to the "choosing the first English word in our translation" event, then perhaps $F = $ 'Twas (if we were translating the French version of "'Twas the Night before Christmas" back into English). As I commented earlier, the outcome of an event is chosen according to a probability distribution. For a die, we would typically assume that the probability of each of the outcomes is $\frac{1}{6}$. For the first word of an English sentence, we would have 50,000 entries in our distribution. By far the most common first word for an English sentence is "The," with a probability of about .03. (About three sentences out of 100 start with "The.") Formally we would write

$$P(F = \text{The}) = .03.$$

(I would warn you that I just made this up, except no scientist in their right mind would say this. Scientists do not make up numbers—we subjectively evaluate them.)

We can also talk about probabilities of two events both happening, as in $P(F = \text{The}, S = \text{violin})$, where S is the event "pick the second word of our English translation" and the comma between them is meant to make you think of a list of things that are all true, so you can think of it as representing "and."

Sometimes two events can be *independent*. Independent events are those that have no effect on one another. For example, if we roll our die before choosing the first word of the translation, we could talk about $P(R = 3, F = \text{Truck})$. When talking about probabilities of two or more events, we use the expression *joint probability*. In particular, if the two events are independent, their joint probability is the product of their individual probabilities

$$P(R = 3, F = \text{The}) = P(R = 3) \times P(F = \text{The}).$$

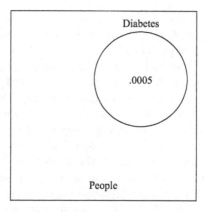

Figure 3.1: Venn diagram of diabetic patients from the universe of people

Another important connective between events is *conditional probabilities*. A patient walks into a doctor's office in want of a diagnosis, D. We know nothing about the patient, so the probability that this patient has, say, diabetes, is the prior probability for this diagnosis, perhaps .0005, and we would write

$$P(D = \text{diabetes}) = .0005.$$

To repeat our example from section 3.1 of great thirst, call this event T. We are now interested in the conditional probability of D *given* thirst T is high or

$$P(D = \text{diabetes} \mid T = \text{high}) = .5.$$

Formally, with a prior probability, we have a universe of people from whom we are choosing, and the question is, what is the probability that in randomly picking a person from that universe we pick a person who happens to have diabetes? Often we use *Venn diagrams* to illustrate these situations. Figure 3.1 shows the subset of people who have diabetes within the universe of people.

When we condition one random event on a second, we are choosing not from a preunderstood universe but rather from the universe defined by the second event—in this case, the universe of people with great thirst. In figure 3.2, we use a Venn diagram again. As we have drawn it, half of the people with great thirst also have diabetes, reflecting our assessment that $P(D = \text{diabetes} \mid T = \text{high}) = 0.5$.

Formally, we define

$$P(D \mid T) = \frac{P(D, T)}{P(T)}.$$

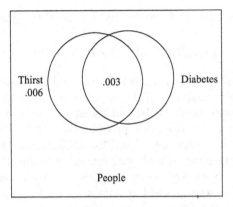

Figure 3.2: Diagram for probability of diabetes given thirst

Repeating what I just said in light of this definition, we are concerned with the people with both diabetes ($D =$ diabetes) and great thirst ($T =$ high) as a fraction of people with just great thirst. You might also note that in this definition I omitted specifying the value of the events. When one does this, it means that the equation holds for *any* outcome values you specify. (Except that $P(T)$ cannot be zero; remember that you are never allowed to divide by zero.)

People can sometimes get confused between the conditional probability $P(A \mid B)$ and the conditional probability with the two events reversed, $P(B \mid A)$. They generally are quite different. Consider the probability of flu given a sore throat. Flu is pretty uncommon, say. .001. If you have a sore throat, that increases the probability, but it is still uncommon, so perhaps the probability of flu given a sore throat is .002. On the other hand, if you have the flu, it is quite likely you have a sore throat. So, reversing the conditioning to probability of a sore throat given the flu might be .5. I am making this up. I know almost nothing about the flu. But if I have this wrong, substitute some other common ailment caused by both a cold and the flu. The probability of the symptom given the flu is much higher than vice versa.

Two more important results before I let you go from this impromptu lesson in probability follow. One is the chain rule:

$$P(A, B, C) = P(A) \times P(B \mid A) \times P(C \mid A, B).$$

It says we can break the probability of three events into three probabilities. We can see this in several ways. Perhaps most intuitively, suppose A, B, and C were independent of one another. Say A is getting heads, B is rolling a 6, and C is drawing an ace of spades. The probability of all three happening is just the product of the three prior probabilities, as shown here:

$$P(A, B, C) = P(A) \times P(B) \times P(C).$$

But if the events are not independent, we need to condition the second two. For example, suppose A, B, and C are choosing the first three words for the next book to be published in England. We guess the first from pretty much the prior probabilities for words. "The" would be high, "Leaves" pretty low. "Antidisestablishmentarianism" (which, when I was in grammar school, was famous as supposedly the longest word in English) very low. Once the first word has been chosen, the probability of the second word is not the prior for that word, but its probability given the first word. In a similar fashion, we get the third word conditioned on the two-word sequence that preceded it.

Another way to see that the chain rule is true is to insert the definition of conditional probability everywhere it appears on the right-hand side of the equation:

$$P(A, B, C) = P(A) \times \frac{P(A, B)}{P(A)} \times \frac{P(A, B, C)}{P(A, B)}.$$

On the right-hand side, all of the denominators cancel out with previous terms, leaving just the left-hand side. (Indeed, the first time I saw this, the equation looked so simple that I had a hard time believing it was ever useful. But it is a mainstay of the use of probability theory in language processing, including machine translation.)

Earlier I said that a probability distribution over several events, such as we see on the left in the previous equation, is called the joint probability. You can think of it as a big table. To take a simple case, suppose we have two events: one is W, tomorrow's weather, with outcomes fair, cloudy, and rain, and the other is my health tomorrow, H, with outcomes good or bad. Their joint probability might be:

	$W = $ fair	$W = $ cloudy	$W = $ rain
$H = $ good	.2	.3	.1
$H = $ bad	.2	.1	.1

First, note that all of the numbers sum to one. This is, of course, dictated by the fact that both events have to have *some* outcome.

Less obvious is that *any* probability dealing with the events W and H can be computed from this table. For example, if you want to compute the probability that I will be healthy tomorrow, you sum over the three probabilities in the top row, being healthy in fair weather, in cloudy weather, and in rain: (.2 + .3 + .1 = .6). This property makes joint probabilities good things to have. Summing joint distributions to obtain more specific ones is called *marginalization*. It is said the name refers to the fact the margins of a page are good places to write down a sum of a row of numbers.

If you have stuck with me so far, congratulations! You are now ready to derive the most famous equation in the theory of probability, Bayes' law:

$$P(A \mid B) = \frac{P(A) \times P(B \mid A)}{P(B)}.$$

The derivation is quite simple:

$$P(A \mid B) = \frac{P(A, B)}{P(B)}$$
$$= \frac{P(A) \times P(B \mid A)}{P(B)}.$$

The first line is just the definition of conditional probability. In the second line, we replaced the numerator using the chain rule.

To better understand the importance of Bayes' law, we slightly rearranged the terms:

$$P(A \mid B) = \frac{P(A, B)}{P(B)}$$
$$= P(A) \times \frac{P(B \mid A)}{P(B)}.$$

Bayes' law tells us how to adjust probabilities when new evidence comes along. It says, in light of the evidence provided by B, the old probability of A, $P(A)$ should be multiplied by

$$\frac{P(B \mid A)}{P(B)}.$$

This is the probability of the new evidence B given A divided by B's prior. This makes sense. Suppose B is a common symptom, like sniffles. Then $P(B)$ might be as high as, say, $\frac{1}{8}$. The point is that even if A always causes B (i.e., $P(B \mid A) = 1$), the presence of sniffles is only going to increase the probability of the disease by a factor of 8. A factor of 8 is not much if A is rare. Conversely, if B is rare, its presence here would have a large influence on our diagnosis of A.

3.4   Discontents with Standard Probabilities

The problem, then, for programs like MYCIN is that classical probability and statistics did not seem to fit their needs. What follows is mostly taken from introductory sections of an article about MYCIN that detail this to justify the use of their new scheme [96]. While I am going to conclude here that they were all wrong, I should emphasize that I am pretty sure I read the article soon after it was published and *I agreed with it completely.* So, I really do not want to cast the first stone, though I suppose it looks like I am doing so.

When one reads mathematical texts, one gets overwhelmed by the requirements before one can calculate probabilities in a realistic domain such as medicine. First, one needs prior probabilities. If you ask a doctor the prior for diabetes, they will look at you blankly. Earlier, as an example, I guessed .005, but I would hardly be surprised if I was too high by a factor of 10, or 50 for that matter. More generally, at the time of MYCIN, there were almost no medical databases with empirically observed probability estimates, and what existed were for very specialized subsets of diseases.

For combinations of findings, the available data are even more scarce. As we noted in our section on probability and statistics, the prior of two events is only the product of their individual priors if the events are conditionally independent, and in any cases we might care about, this is hardly ever true. (That is to say, the probability of having a cough can be considered conditionally independent of diabetes, but in the cases we care about, symptoms of related diseases, this is rare.)

Next, there seemed to be issues about what probabilistic information *means.* To paraphrase an example from the MYCIN paper, if the probability of diabetes given thirst is 0.6, we would be inclined to say that excess thirst supports a diagnosis of diabetes to the degree 0.6. But does this mean no excess thirst supports the diagnosis of diabetes to degree 0.4 ($=1-.6$) (reasoning here that the probability of X plus the probability of not X must sum to one)? Common sense would seem to say "no," but the law of the excluded middle seems to say "yes."

In reaction to these issues, MYCIN has a system where propositions have two numbers associated with them: one is a degree of belief, the other, disbelief. This handles the last issue raised earlier. MYCIN also restricts the use of such numbers to rules of the form, "if you see evidence X, conclude disease Y with certainty Z," or "conclude not Y with certainty Z." These seemed to be the propositions with which doctors were most comfortable when asked about degrees of belief. And, of course, MYCIN's numbers were forthrightly doctors' belief estimates, and as such, issues of formal tables of probabilities or formal independence assumptions were thought not to apply.

3.5   Alternatives to Standard Probability

By the mid-1980s, there was no consensus on what to do about uncertainty in AI. MYCIN and a few others programs specifically aimed at diagnosis issues introduced their own newly designed systems. Other methods were designed with the goal of completely replacing probability and statistics, of which two deserve particular mention.

One was *fuzzy logic*. From the viewpoint of fuzzy logic, the problem of uncertainty arose from the fact that standard logic is *Boolean*. In Boolean logic, a proposition could be either true or false, or, more technically, propositions had two *truth values*. Fuzzy logic was introduced to AI by University of California at Berkeley professor Lotfi Zadeh [114]. (Previous research on this topic went under the name *infinite-valued logic*, but it was Zadeh who championed its use in AI, and fuzzy logic was the term Zadeh used.) For example, we might now say

(size ball-2 large) 0.6,

where the associated number, 0.6, states that we have some, but not complete, belief that ball-2 is large. The intent is to replace both Boolean logic and standard statistics and probability with a logic that includes numeric levels of belief.

Zadeh had been publishing work on fuzzy logic since the 1960s, but it received renewed interest in the '80s for several reasons, the foremost being the general confusion with regard to uncertainty. However, fuzzy logic had several other things going for it. It was comparatively simple. The basic idea seemed both obvious and reasonable, and the rules for manipulating degrees of belief were less complicated than for the alternatives.

Another big selling point was its popularity in Japan. If you follow the news anywhere near as faithfully as I do, you are aware that the United States sees itself in competition with China these days. I say "these days" because in my lifetime, there have been periodic waves of such concerns. By "competition" here, I do not necessarily mean military competition, though it could mean that as well. Rather, I am talking about issues concerning the basic organization of society and how that affects the economy, quality of life, and that sort of thing. In the '50s and '60s, the economies of communist countries were growing at a much faster pace than the US economy, and the fear was that a more forceful, top-down organization might inevitably outpace our more bottom-up, capitalist one. As it turned out, the observation was true, but the conclusion was wrong. Rather, the United States started adopting technological and organizational improvements much earlier than the Soviet Union, which was picking the low-hanging fruit.

The point is, in the '80s and '90s, Japan was on a similar roll. Japanese society does not look much like Russia's, but it is used to much more government control than society in the United States. Furthermore, the Japanese are much more inclined toward sacrificing individuality in support of collective effort. We worried again, and it made headlines when one Japanese company bought Rockefeller Center, the great symbol of US capitalism. I remember at least one action film where the bad guys were some Japanese corporation. So, when a Japanese government agency adopted fuzzy logic for a new funding project, the *New York Times* published an article on how the Japanese were quick to pick up and run with American ideas. The Japanese economy tanked shortly thereafter. (But you can still buy several brands of fuzzy-logic rice cookers. Fuzzy logic was one of the few AI mistakes that I *did not* fall into, so I can look down on those that did. Thus, the line about rice cookers is to be read with disdain in my voice. More seriously, though, I never found out if the Japanese researchers ever did use the scheme for anything useful.)

A third proposal for handling uncertainty was *Dempster–Shafer theory* [95]. As with the MYCIN system, it, too, associated two values with each proposition: an index of belief and one of disbelief. However, these values did not need to sum to one. Instead, an **any** category picks up the slack. If **any** is not zero, it means a degree of puzzlement exists about the proposition and its negation. Dempster and Shafer were particularly worried about setting priors in cases of virtually no information, such as "there is extra terrestrial intelligent life in the universe." Here, they might put all the weight on **any**. Dempster–Shafter was complicated, and while it had its supporters among the more mathematically inclined, I am not aware of any working program that used it.

3.6   Bayesian Networks

What happened next is, to my mind, a great testament to how science works at its best. At the 1984 meeting of the Association for the Advancement of Artificial Intelligence, a panel discussed the issue of handling uncertainty. By the end of the session, everyone was unhappy in that they felt that the others were simply not recognizing the force of their arguments. Most unhappy was Judea Pearl, the sole defender of classic probability and statistics, albeit in its Bayesian guise. So, all the panel speakers got together and started a new conference named Uncertainty in Artificial Intelligence, which was open to papers from all these approaches. Further, the paper-submission committee was designed to have representatives from all the competing camps. I myself was now concerned with these problems (as they related to understanding text) and started submitting papers to the conference, starting about 1990. The conference is still alive and active but not serving its original purpose. Judea Pearl published his book, *Probabilistic Reasoning in Intelligent Systems* [78], in 1988, and his once lonely defense of standard probability and statistics became the AI-wide solution within a few years.

What about the aforementioned objections to standard probability? Some are just objections to frequentist statistics and are "solved" by becoming a Bayesian. The MYCIN researchers stated that their numbers were not probabilities but, rather, statements of physician's beliefs. A Bayesian would simply reply that they could be both.

Some of the difficulties are just confusion due to the use of intuitive English when discussing abstract concepts like probability. Conditional probabilities can be particularly slippery. If T is thirst level and D is the diagnosis, then $P(D = \text{diabetes} \mid T = \text{high}) = .6$ is expressed in English as "given excess thirst, the probability of diabetes is 0.6." From there, we move on to say "excess thirst supports a diagnosis of diabetes to the degree .6." But that is only an informal paraphrase, and if it leads you to conclude that "not having excess thirst supports diabetes to degree .4," then it is a bad, misleading paraphrase. Why? Because if we actually work through the conditional probability math, the fact that the patient does not have excess thirst *lowers* the probability of that diagnosis, as it should. (That's true if one assumes, as is reasonable, that at any given time, more people are thirsty than have diabetes.)

Another objection is that doctors are uncomfortable dealing with probabilities. In general, everybody is. The psychologists David Kahneman and Amos Tversky have shown that people can be *terrible* at estimating probabilities, to the degree that they make statements that contradict the axioms of probability theory [47]. Their classic example is:

> Linda is 31 years old, single, outspoken, and very bright. She majored in philosophy. As a student, she was deeply concerned with issues of discrimination and social justice, and also participated in anti-nuclear demonstrations. Which is more probable?
>
> 1. Linda is a bank teller (A).
>
> 2. Linda is a bank teller (A) and is active in the feminist movement (B).

Many people choose the second despite the fact that the probability $P(A, B)$ can **never** be higher than the probability $P(A)$. Remember the Venn diagrams we showed earlier. The probability of the joint statement is proportional to the intersection of, in this case, bank tellers and feminists, which cannot be larger than the group of all bank tellers. Instead, what Kahneman and Tversky found is that when asked to estimate probabilities, people seem to respond with estimates of *typicality*. A "typical" bank teller might not sound like someone interested in social justice, but make her a feminist and it sounds more likely. This is now known as the "Linda problem."

So, people make mistakes because they do not understand probabilities. But probability is normative—if you can use it, you should, because you cannot do better. Like many of us in AI, I give some weight to the statement, "but people do not do it that way." But if I had my druthers, I would rather have the people or programs diagnosing me use normative methods.

As for the lack of the probabilities necessary for actually computing conditional diagnosis probabilities, the current, quite good answer is provided by *Bayes nets*, or equivalently, *Bayesian networks*. Bayes nets are the work of many contributors and have received other names, such as *belief networks* or *decision networks*. "Bayesian networks" is the name Judea Pearl used. I briefly mentioned Pearl earlier when I said that he was the sole defender of standard probability in a panel session on uncertainty in AI. (Pearl has many, many important contributions to AI and beyond. He is also largely responsible for our current understanding of causality and its mathematics.)

A Bayes net is a *directed graph*—a set of nodes (also called vertices) connected by arrows. There may not be any directed cycles. Each node represents an event, where we are using "event" as it is used in probability. However, Bayes nets are deliberately meant to have both a probabilistic and a more intuitive causal interpretation. Figure 3.3 is (one version of) a standard in the literature, the "earthquake" example. If you follow the arrows, you see that either an earthquake or a burglary can cause my house alarm to go off, which in turn can cause my neighbors to call me (assuming I am at work, say). The prohibition against directed cycles stems from the interpretation of edges as statements about causality plus the assumption that the network represents the state of the world at one time.

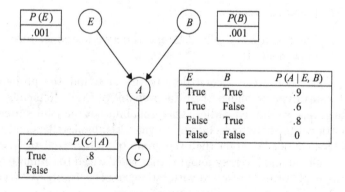

Figure 3.3: Bayes net

The probabilistic interpretation is that each node is a random event and has associated with it a probability distribution for each event outcome given the parents of the node. So, the node C (receive a call from your neighbor) has associated with it $P(C|A)$—C is conditioned on one event since it has one immediate parent in the network.

Bayes nets are used to compute conditional probabilities by "setting" the values of nodes for which we know the outcomes. We can then compute the posterior probabilities of the others. So, if, say, I received a call, and I want to compute the probability that I have been burgled, then I want to compute $P(B = \text{True} \mid C = \text{True})$, so I "set" C to "True" and recompute the newly conditionalized probabilities for the rest of the nodes. It is possible to do this computation despite the fact that we have specified only a handful of all the probabilities you might think up involving the four events. We can do this because several independence assumptions are built into a Bayes net.

Remember that Bayes nets deliberately encourage a causal interpretation. The fact that no downward arrow leads into node E means that, as far as this network is concerned, earthquakes have no causes. A similar analysis applies to burglaries. This means these two probabilities are independent of each other, as shown here:

$$P(E, B) = P(E) \times P(B).$$

Another independence assumption is that once you know a node's immediate parents, the value of the node is independent of any other predecessor for the node, where a predecessor node is one with a directed path to the node. Given these independence assumptions, we can show that the joint probability of all the nodes in a Bayes net can be expressed with just the individual probabilities for each of the nodes in the network. For the network in figure 3.3, this is:

$$P(E, B, A, C) = P(E) \times P(B) \times P(A \mid E, B) \times P(C \mid A).$$

To see that this set of distributions is sufficient, consider the distribution $P(E, B, A, C)$. As you may remember, when we do not include outcomes for the events, we mean this to represent all possible combinations. So, for example, the probability for $E = \text{True}$, $B = \text{False}$, $A = \text{True}$, and $C = \text{True}$ would be for the situation where there was an earthquake, no robbery, the alarm sounded, and there was a call. (If you are finding this discussion tough sledding, you can skip this and the next paragraph if you like.)

Next, we can use the chain rule to break this distribution up into pieces, as

$$P(E, B, A, C) = P(E) \times P(B \mid E) \times P(A \mid E, B) \times P(C \mid A).$$

These are exactly the component probabilities that we claim are sufficient. Furthermore, while in the general case, large Bayes nets can be
expensive to evaluate, more often than not, things do not get that bad.
Another neat feature is that if you specify just the probabilities required
by a Bayes net, you are guaranteed no contradictions exist. For example,
it cannot be the case that $P(A) = 0.5$, $P(B \mid A) = 0.5$ and $P(B) = 0.1$. If
half the world has a cold, and half of those with a cold have sniffles,
it cannot be that only one in ten have sniffles. This is an easy case,
but conflicting probabilities can be quite difficult to detect. This was
another objection to standard probability and statistics. Also, because
of their causal interpretations, people find the numbers reasonably easy
to estimate (or make up).

As more people adopted Bayes nets, the Uncertainty in Artificial
Intelligence conference lost the need to be a place to debate alternative
approaches, and it now concentrates on research with classical probability at its base. I myself became a convert in about 1990 and wrote a
guide [15] titled "Bayesian Networks without Tears." Geoff Hinton (an
important person in the current deep learning surge) suggested that I
title it "Pearl for Swine." Unfortunately, the editor of *AI Magazine* had
no sense of humor and would not let me use it on the grounds that I was
calling his readership pigs.

Bayes nets are in many respects a high point of symbolic AI. They
elegantly solve numerous problems of representing and reasoning about
uncertainty. However, they are clearly wedded to the symbol system
hypothesis since the objects that are judged uncertain are propositions.
The Uncertainty in AI conferences are ongoing and have workshops that
concentrate on applications of the technology, so they are in practical use.
However, as we see by the end of this story, this dependence prevents
them from participating in the current waves of progress and excitement
that deservedly grip our field.

4

Chess (1965–1997)

Our history is up to roughly 1985, and we have been talking about expert systems and related issues. AI is about to move toward machine learning as well as become more comfortable with numbers (as we have seen with Bayes nets). However, 1985 was also a significant year in AI chess. While chess has been a major topic for AI, we have ignored games since our discussion of checkers because its history is largely separate. I guess this is a polite way to say that AI chess is somewhat of a side-show. Nevertheless, it is a fun topic, and any history of AI would be less for ignoring it. It is also the case that chess was "solved" (chess programs surpassing people) in a surprising fashion.

Checkers, chess, and Go are all two person games of *perfect information*. By "perfect information," we mean that both players know everything there is to know about the current and all possible future states of the game. This follows from knowing the positions of pieces on the game board since the game contains no element of chance or luck. In principle, you could try out all possible moves, countermoves, counter-countermoves, and so on and select an optimal strategy. In practice, this is not possible because too many alternatives exist.

Generally, games are thought to be good testbeds for AI ideas because of the ease of judging which of several competing programs is the best—have them play each other. Perfect knowledge games are even better because they have no element of chance to confuse the evaluation. Thus, it is not just an accident of history that this book's first example of an AI program was one to play checkers or that one of the early (and typically optimistic) predictions of what AI would accomplish was by Herbert Simon (one of the Dartmouth Conference attendees) that in ten years, an AI program would beat the world champion in chess. This forecast was made in 1957. Samuel's checkers program was playing very good checkers by then, but in the West, checkers is considered a game for children while the primary intellectual adult game is chess.

I remember John McCarthy (another Dartmouth pioneer) giving a lecture at the MIT AI lab at some point in 1968 and mentioning Simon's prediction, which had obviously failed to materialize by that point. McCarthy pointed out that although Simon was wrong, he had clearly put some thought into his prediction. In fact, according to my memory of the talk, McCarthy noted that Simon had made not one but three predictions for what AI would have accomplished by 1967. The first was about chess, the second was that an AI program would compose a good piece of music, and the third, that an AI program would prove a significant mathematical theorem. McCarthy said that Simon had clearly looked for areas that produce prodigies. McCarthy noted that prodigies in these fields could make major achievements with comparatively little experience. This made it more likely that computers might do the same.

4.1   Chess Program Organization

Going back to our discussion of Samuel's checkers program, it consists of three subprograms, of which the first is the plausible move generator. At any position in the game, the plausible move generator selects from among the legal moves those that are more likely to lead to a win. Second, the minimax search projects the consequences of each plausible move. If I consider all my opponent's possible moves in response to one of mine, this gives me a more sound estimate of the quality of that choice. This would be a one-*ply* look-ahead—a ply is a search of several alternative moves for me and my opponent. A program can do a multiply search, say, five ply—five moves and countermoves into the future. The number five is a choice made by the programer, but the deeper we can search without running out of time, the better. How deep we can look ahead also critically depends on how many moves are returned by the plausible move generator. The more it returns, the less deep we can look ahead in the game before the total number of positions under consideration overwhelms the machine. For example, suppose in each board position I consider four possible moves. If I consider four possible responses by my opponent to the four moves I might make, I will have considered $4 \times 4 = 16$ board positions. Furthermore, if I continue on in this fashion (considering four of my possibilities to his sixteen boards), it would require $4^3 = 4 \times 4 \times 4 = 64$ positions. This kind of expanding tree is illustrated in figure 4.1.

Suppose, however, that our plausible move generator returned not four positions, but ten. Suddenly, the program is going to be looking at

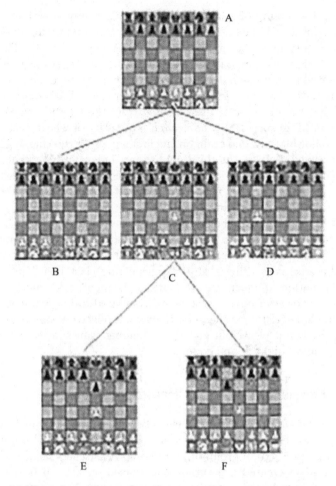

Figure 4.1: Board positions after three possible moves and two responses to one of them

10^3 positions—that is, 1,000. If we were looking three ply ahead (three move/countermove pairs), the difference between considering four plausible moves versus ten is even more dramatic because we raise four or ten to the sixth power: $4^6 = 4{,}096$ and $10^6 = 100{,}000$. For this reason, the plausible move generator was thought to be the most important piece of a chess program. This belief turned out to be completely wrong.

The general point is that the number of board positions considered (n) grows *exponentially* in the depth of our search, d. If our move generator produced, say, four moves when presented with a board position,

checking the quality of a move by looking only at my opponent's immediate response would examine four positions. (I am repeating myself, but
this is important.) The number of moves for depth d is 4^d. This is an
exponential of d, where d is the *exponent*. In computer science, seeing an
exponential is an immediate warning not to make the exponent too large.
(Indian mathematicians knew this a long time ago. Early Indian legend
has it that the inventor of chess wanted from the king a reward for the
invention. More specifically, he wanted a quantity of wheat necessary to
do the following: put one grain on the first square of the chessboard, two
on the second, four on the third, and so on. This is also an exponential,
2^n. Since there are sixty-four squares on a chess-board, the last square
would require $2^{64} = 1.8446744 \cdot 10^{19}$ grains of wheat, which is about the
number of atoms in the universe.)

When we have reached our self-imposed search limit, we use the board
evaluator to estimate how good/bad the position is for me or my opponent. This piece of the program is also called the *static board evaluation* because it is using a static (unchanging) last board position as
the information to inform the evaluation. We mentioned that how many
more/fewer pieces I have on the board compared to my opponent would
be a simple heuristic for checkers. In chess, a relatively simple evaluator
would also count pieces but give them different values: a pawn would be
worth 1, and knight 2, and so forth.

4.2   Bernstein's Chess Program

Early on the major constraint on chess programs was how much memory
there was for the game-tree search and how fast we could compute the
moves. In the 1950s, a group of atom-bomb researchers headed by John
von Neumann created a program that played on a 6×6 board. Chess
on a 6×6 board is played without bishops. (Von Neumann is another
towering figure in CS history, having invented the so-called *von Neumann
architecture*, the basis of computer architectures used to this day.)

The first full-board chess program was that of Alex Bernstein [5].
Bernstein was working at IBM and thinking about writing a chess program. He was invited to the Dartmouth Conference, where he met John
McCarthy. Their conversations led McCarthy to the next big improvement in two-person games—*alpha-beta search*. We have already mentioned minimax search. Alpha-beta search is an improvement in that
it retains the optimal performance guarantees provided by minimax but
searches many fewer game states. The basic idea is this: Suppose you have
four possible moves, and you have found that if the program chooses the
first of these, it results in a board position worth +5. You now consider

move 2. Your plausible move generator suggests four moves for the opponent to your move 2, and the first one limits you to +3. At that point, you should not explore move 2 any further, since with move 1 you can get +5, but we now know that with move 2 you do not get anything better than +3 since, unless there is something even worse, your opponent will choose the response that limits you to +3.

Bernstein's program using alpha-beta search searched four ply. It had a chess rating of about 500, a complete beginner. For comparison, 1,000 is a beginning tournament player, 2,000 is a grandmaster (an internationally ranked player), and someone rated 2,500 could be the world champion.

4.3   Mac Hack and Chess 4.0

The first tournament-level chess program was Mac Hack. It was developed at MIT by Richard Greenblatt [31]. It initially had a rating of 1,000 and, with its rough edges smoothed over, routinely, reached 1,200 in tournament play. Mac Hack had a few innovations, but mostly it was Greenblatt's master programming ability that made the difference. It had few, if any, significant bugs, and Greenblatt programmed the entire thing in assembly language, allowing the program to explore more deeply than its computer competitors. (Assembly language is the native language for a particular brand of computer and, as such, is trivially translated into 0s and 1s. My programing for the Argonne National Laboratory physics experiment was in assembly language. When we talk of an assembly language, we are contrasting it with a *higher-level* language—one designed to make programming easier but at some expense in efficiency.)

When I got to MIT in 1968, I went along with Greenblatt to a tournament in which Mac Hack was entered. By this point, this sort of thing was routine. It was just a matter of plugging in the teletype and someone typing the moves to the machine. However, I do remember Greenblatt telling me that occasionally a significantly stronger player would lose to the program because they were so freaked out by the idea of playing a machine. Remember, this was 1967.

The next big advance was the program Chess 4.0 by Larry Atkin and David Slate at Northwestern [99]. This was the first tournament-level chess program to give up on improving the plausible move generator and instead go with full-width search (which just means not restricting next moves to some subset of all the legal moves). Atkin and Slate did not set out to rid chess programs of plausible move generators but ran out of time for that year's chess program tournament, so they entered a program without one. It did not win but came close, and in a few years, almost all programs had given up on move generators that tried to incorporate any significant knowledge of chess. This was 1973.

4.4 Deep Blue

Then in 1985, Feng-hsiung Hsu, a PhD student at CMU, began his work on developing special-purpose chess-playing hardware called ChipTest [39]. This hardware could compute possible moves very quickly, resulting in software that could search much more deeply in the game tree. This led to the Deep Thought program in 1988. A year later, Hsu received his PhD and went to IBM. At IBM, he and his group improved Deep Thought and renamed it Deep Blue [40]. (IBM is affectionately known as "Big Blue.") In 1997, Deep Blue effectively ended the interest of the AI community in the game of chess by beating Garry Kasparov, the reigning world champion chess player, thus finally, and belatedly, fulfilling a prophesy made early in the history of the field that computers would soon be the world champion in chess.

(There is some debate about the degree to which the version of Deep Blue that beat Kasparov was his superior. Kasparov lost the final game of the match when he made an elementary error in the opening by making a move that was well known to be unwise. Many in the chess community felt that, as it was the last game, Kasparov was simply tired, a problem that did not affect Deep Blue. However, within a year or two, there could be no doubt that the era of human domination in chess was over.)

A paragraph earlier, I said that Kasparov's defeat "effectively ended" AI's interest in chess. What had become clear was that while so-called brute-force methods (building faster hardware so that we could search ever larger portions of the game tree) would work for chess, this was only because of chess's special place in the hierarchy of game difficulty— just difficult enough that, without the special-purpose hardware, human chess masters could still beat machines but just easy enough to push computers over the boundary. Thus, still more difficult games, such as Go, the intellectual game of choice in Asia, were not affected by this hardware revolution. In chapter 9, we consider Go in more detail and how very different AI methods would lead to its conquest by machines.

5

Computer Vision (1970–2000)

5.1 Hubel and Wiesel

As briefly mentioned in the preface, computer vision is the problem of recognizing and locating the objects in a scene. The input to a computer vision program usually is the light intensities at a (typically large) number of points in the image, called pixels. In the early days of AI, the ease with which humans interpret visual scenes led researchers to vastly underestimate the difficulty of the problem. At MIT, Richard Greenblatt was hired to lead a group of undergraduates for a summer project to solve vision. This was while Greenblatt was still an undergraduate himself and before he did his work on chess. Needless to say, they made no real progress, and Greenblatt never worked on vision again.

In fact, the most important progress in computer vision during that period was made by two neurobiology researchers, David Hubel and Torsten Wiesel [41]. They inserted electrodes into a cat's early *visual cortex*, the part of the brain that receives information directly from the eyes and starts the process of interpreting the light intensities. The idea was to see what sort of images would cause the neurons near the probe to fire. Initially, they had no luck, as they could not see any connection between the scenes and the neuronal firing patterns.

Actually, I was assuming here that the reader already knows some neuroanatomy, but perhaps you do not. If you do, you can skip to the next paragraph. For the rest of you (and me, of course), the brain is made up of individual cells called *neurons* that are connected to each other at *synapses.* Information is processed by electric currents that pass between the neurons (at synapses). We say that a neuron *fires* when it generates a current, typically because neurons that connect to it have recently fired. Learning takes place by modifying the synapses so that the electric signal can pass more easily to the receiving synapse (or less easily). Neurons can fire for no particular reason, so a firing by itself is not particularly meaningful. Rather, what counts is the *rate* of the firing. Also, neurons are more plentiful at the *fovea*, in the center of the retina, so animals shift their view to get a more complete perception of the scene.

Coming back to Hubel and Wiesel, they noticed no connection between what a cat was shown and which neurons fired, until one day they saw that a group of neurons fired not at the scene but when the projector was going between one scene and the next. The projector showed a line or two somewhere on the screen when it was between images.

The rest, as the saying goes, is history. Further experimentation showed that these particular neurons were firing whenever a line appeared at a particular orientation and place. Furthermore, if a particular set of neurons responded to, say, vertical lines (90°), then its neighbors to the right would respond to orientations of 95°, and those to left at 87° or some such. This suggested, quite graphically, that the brain starts scene recognition by first finding lines and does so globally, before it does anything else, and not sequentially as it would if it were line tracing—looking for places of maximum intensity change and then tracing outward, again, presumably in the direction of maximum change.

Naturally, it took some time for the implications of this work to influence computational research. I was hired to work at the MIT AI lab the summer before I started as a student at MIT. Thomas Binford, an early AI vision pioneer, took me on for the summer to work on line finding. I was completely ignorant of Hubel and Wiesel's work, and if Binford knew of it, I have no memory of him mentioning it to me. My line finder worked poorly, and I, too, was introduced to the illusion of the simplicity of visual processing. (It may have also contributed to my early antinumeracy, mentioned in chapter 1.)

5.2   Convolution

When the implications of Hubel and Wiesel's work did sink in, the result was a general adoption of *convolutional methods*. When we speak of convolution in the context of vision research, we have in mind the following idea. Suppose a small portion of our image has light intensities that look like those shown in figure 5.1. Each number represents the light intensity of a single pixel, and this corresponds to a piece of an image where a line

$$
\begin{array}{ccccc}
5 & 0 & 0 & 0 & 0 \\
5 & 5 & 0 & 0 & 0 \\
5 & 5 & 0 & 0 & 0 \\
5 & 5 & 5 & 5 & 0 \\
5 & 5 & 5 & 5 & 0
\end{array}
$$

Figure 5.1: Light intensities for a diagonal change in light intensity

occurs due to an abrupt transition from light to dark. Furthermore, the line goes diagonally across the image, from left to right.

We are also assuming we are dealing with black-and-white images, so we need only one number to represent the level of light. If we had a color image, we would need three numbers at each point, one each for the amount of red light, blue light, and green light. This would be an RGB representation of the color image, where, of course, RGB stands for red, green, and blue. Any color can be created by a mixture of these three. (Not obvious, but true. One thing most of us are confused by is our experience of mixing paint pigments. With paints, we want a surface that *reflects* a particular color, and if we mix red, blue, and green in equal quantities, we get black, since the surface does not reflect any of the colors. When we mix light, it works almost in the opposite way. The mixture of red, blue, and green lights is white light.) Getting back to convolution, consider the following small patch of numbers:

$$
\begin{array}{rrr}
-1 & -1 & -1 \\
1 & 1 & -1 \\
1 & 1 & 1
\end{array} .
$$

We now do something that at first appears quite strange. We take the patch and center it over a 3×3 piece of the image and multiply each number in the patch by the corresponding number in the image. For example, if we choose the upper-left-hand side of the image in figure 5.1, we are multiplying these two groups:

$$
\begin{array}{rrr@{\qquad}ccc}
-1 & -1 & -1 & 0 & 0 & 0 \\
1 & 1 & -1 & 5 & 5 & 0 \\
1 & 1 & 1 & 5 & 5 & 0
\end{array} .
$$

For example, in the first row, we would multiply each -1 times 0. In the second row, we get 1×5, 1×5, -1×0. After doing this, we get nine products, and we add them all up. The first row contributes 0, the second adds up to $5 + 5 + 0 = 10$, and the third to 10 as well, giving us 20.

To begin to see the point of all this, contrast it with what we would get if the image had no light intensity difference at the point in question.

Each number would be the same, perhaps all 5s. When we do the same multiplications and additions, we get −5 for each pixel that corresponds to a −1 in the patch, and 5 when there is a 1. Since there are four −1s, and five 1s, the total would be 5. More generally, the closer the image region is to a diagonal line from left to right, with the brighter part on the left, the larger the resulting multiplications plus additions. We have, in other words, created a diagonal line detector, just like one of the neurons that Hubel and Wiesel found in the cat. Furthermore, by changing the size of the patch and the numbers in it, we can find lines of any angle we want. Multiplying a variety of patches at each point will detect where there are lines, and what their angles are, for the entire image. This process is called *convolution*, and the technical term for the patch of numbers is a *kernel*.

5.3   Stereo Vision

Most visual scenes are two-dimensional recordings of originally three-dimensional scenes. This suggests that one way to move beyond line finding, but not yet to the point of identifying objects, is to reconstruct the 3D placement of objects or planes in the scene. This is called depth perception. Unfortunately, this is difficult, as the standard TV-camera input of a scene contains no such information.

One solution to this problem is to use two or more cameras at separate but not too distant locations. (Equivalently, if the scene is static, one may move the same camera and take a second image.) Figure 5.2 shows two images of a statue. One can best see that the viewpoint has shifted by

Figure 5.2: Side-by-side stereo images

looking at either the extreme left or right sides. For example, a statue appears to the left of the one centered in the image. You can see a slice of it in the left-hand-side image that does not make it into the right-hand version. Conversely, the center statue is holding a book whose right-hand tip just barely makes it into the right-hand picture, but it more comfortably fits into the left. (With some practice, you might be able to "fuse" these two images by crossing your eyes so your right eye sees the left image and vice versa.)

If one can recognize the same point in the 3D world in the two versions, and you know the (a) distance between the two cameras and (b) the camera focal-length, then (given certain simplifying assumptions) it is possible to turn the positions of the same real-world point in the two images into a good estimate of the distance from your eyes to the object. The basic geometry of the situation is shown in figure 5.3. Here point A represents the location of the target object, while B and D are the positions of the two cameras. The various lines are there to create the geometry that will allow us to compute the distance from the cameras to the target.

If you have any interest in photography, you may be familiar with the term *focal length* from seeing it used to describe the magnifying power of a lens. Strictly speaking, it is the distance between a lens and point beyond the lens where the image is in focus. In figure 5.3, this is the distance B-F (or D-G). Since this and the intercamera length are intrinsic properties of our cameras, we assume they are known. The distance we want to deduce is that from the object to the camera—the length of line AC.

The key point is that the two triangles in the upper part of the image (ABC and ACD) are proportional to those in the lower part (BEF and DGH). Thus, one of the equations expressing this proportionality is

$$\frac{AC}{BD} = \frac{BF}{EF + GH}.$$

It expresses AC in terms of distances we know, focal length, intercamera distance, and the positions E and H on the two images. A touch of first-year high school algebra and we are done.

Furthermore, if two images do better than one, even more images are better still. I remember enjoying a talk by Herbert Simon of CMU on recent vision work there. One of the projects analyzed scenes using multiple photographs by Takeo Kanade [35]. This was about 1985, and the venue was the International Joint Conference on AI (IJCAI), so I was still routinely attending talks on current research in computer vision.

One possible caveat to the previous method for determining depth is that we need to recognize the points in two images that correspond to the same location in 3D space. If you or I were asked to do this for points

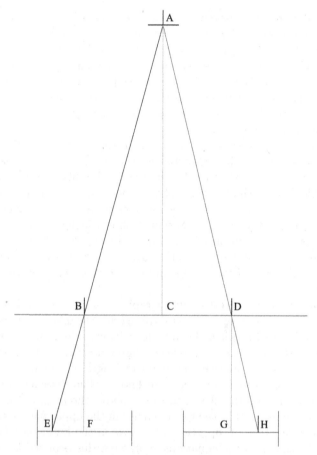

Figure 5.3: Geometric relations to compute distance from two cameras

in figure 5.3, we would locate a distinctive point on an object in one of the images, say, the tip of the book that is just visible on the right-hand side image, locate the book on the left-hand image, and find the same point. Unfortunately, this seems to imply that we need to identify objects before we can determine depth.

Fortunately, this is not the case, as shown by *random-dot stereograms*. As seen in figure 5.4, random-dot stereograms are not much to look at— two images of random dots. However, invisible to a casual observer, a square patch of the dots in one image has been moved a small amount to the left. If you show the two images separately to a person's left and right eyes, they see this patch as above or below the plane of the image. (Again, with practice, you could cross your eyes and see the square

Figure 5.4: Random-dot stereogram

as a hole in the background or "defocus" your eyes and see the patch hover above the background.) Random-dot stereograms were invented by the psychologist/neurophysiologist Béla Julesz and popularized in his book [44].

Of course, the point here is that depth perception must not be dependent on object recognition, since these stereograms have no objects. Rather it must be possible to correlate left and right images just through coincident blobs. Some early influential work on algorithms for doing so, in particular, suggesting that one matches edge patches (e.g., as detected by convolution), was done by David Marr and Tomaso Poggio [62].

5.4 Visual Features

People can, and do, interpret scenes quite well without stereo information, and almost none of the images online have information that would allow binocular depth perception. Thus, most work aims at object recognition (and localization) directly from a single image. A large fraction of this work has been aimed at finding visual features that aid this process. A *feature* is a distinctive attribute, and a *visual* feature is one that helps us identify what is going on in an image.

SIFT features were created in 1999 by David Lowe of the University of British Columbia. SIFT, which stands for "scale-invariant feature transform," has been used for many purposes but primarily for object recognition [58]. Figure 5.5 shows how SIFT features aid in this process.

Suppose we want to recognize, among other things, stuffed bears, old-fashioned telephone sets, and jogging shoes. We start by taking reference photographs of the items and identifying their SIFT features—points where the images show light intensity changes that we expect will be somewhat indicative of the object in question. These points are seen best in the bottom image as small white squares imposed on the image,

Figure 5.5: SIFT features at work

along with the outline indicating where the, say, shoe would be if the SIFT points are indeed from that object. In the middle image, we see a case where some SIFT shoe feature points are misidentified, and a shoe is placed in an incorrect location.

We create specific SIFT features (based upon a particular image) by first *smoothing* the image. That is, we replace, say, the nine pixels in a 3×3 patch by a single pixel that is their average. We do this repeatedly, thereby creating versions of the image at five or six scales. We create a SIFT feature when we find a point in the image that has greater illumination than (a) the eight points around it (at a specific level of smoothing), (b) the corresponding positions in the more smoothed image, and (c) the corresponding positions in the less smoothed image. By requiring that the position be brighter at several levels of smoothing, we are making the feature *scale invariant*. In mathematics, we say that some value is *invariant* under some operation if its value does not change when that operation is performed. In this case, the operation is a change of scale. (We smoothed a pixel position that used to represent the light at a 1×1 mm patch for a 3×3 mm patch.)

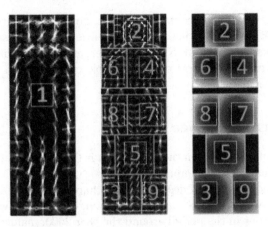

Figure 5.6: HOG features and the deformable parts model

We also make SIFT features rotation invariant. We do this by assigning each possible feature a "standard" orientation. Normally, when we think about feature orientation (for e.g., if we are tracking a line), we care about the orientation of one feature compared to another—are they at right angles?—for example. SIFT features are meant to pick up not lines but, rather, blobs, and the orientation of one SIFT feature only needs to be made the same in all occurrences of the same feature so that we, in fact, recognize them as the same. When sufficient overlap of individual SIFT features exists in our new image with that in the model of a particular object (e.g., teddy bear), we hypothesize that the object is present in the image.

A second popular type of visual features to aid in object recognition is the *HOG* feature put forward by Navneet Dalal and Bill Triggs in 2005 at INRIA, the French National Institute for Research in Computer Science and Automation [21]. HOG stands for *histogram of gradients*. A *gradient* is simply a change in value—here, a change in light intensity. Consider a "medium-size" patch of an image. In the first HOG paper, the researchers used street scenes and wanted to identify pedestrians. See the left-hand section of figure 5.6. A *histogram* is a chart showing the frequency of some quantity, in this case, how often we see light intensity gradients at different orientations. The left-hand portion of the figure shows that the majority of the lines are vertical, with a few diagonal, and almost no horizontal. The idea is that this distribution over orientations is typical for people, similar for, say, fire hydrants, but quite different for dogs or windows. We learn these models from sets of images, some for training the models and a distinct set for testing. The classification processes (going from HOG features to a decision—person/not-person) uses a support-vector machine. (Support-vector machines or SVMs are classifiers, sort of like perceptrons, in that, given some features like the relative quantities of lines in various orientations in this case, they learn weights to classify the input. SVMs, however, are typically more accurate than perceptrons at the cost of considerable complexity in the learning algorithm.)

5.5   Learning to See

The use of visual features in the last section is this chapter's first mention of machine learning in computer vision. As we see when we return to computer vision as part of deep learning (chapter 8), these were not the earliest uses of ML. There was some amazingly prescient work that we discuss later, but in the period around the year 2000, mainstream object recognition was dominated by the work discussed in this chapter, and this early work—gradient detection, or depth perception, or blocks-world block reconstruction—had no ML.

The work on SIFT features changed that, if only by a small amount. Remember, to use SIFT features, we start by extracting them automatically from reference images, as illustrated in figure 5.5. In that work, there were thirty-two reference images, so the program could identify thirty-two types of objects. The test images were transforms of the reference images, where the transforms included rotation, luminosity, and so on. Obviously, before the program processed the reference images, it was incapable of finding the transformed objects, and afterward, it was, so the program exhibited a simplistic form of learning.

A second way that machine learning was entering computer vision work was the introduction of standardized data sets for the training and testing of research programs, in this case the *PASCAL Visual Object Classes (VOC) Challenge* created by Mark Everingham of University of Leeds and colleagues [24]. Each year, the VOC group issued a collection of open-source images along with two labels for each image. The *identification* label specifies what is in the image. In the first year (2005), there were four possibilities: person, bicycle, motorbike, and car. There is also a *detection* label that specified the coordinates of a *bounding box* for objects identified. See figure 5.7. Each year of the competition had a new set of images, and the number of object types increased to twenty by 2007, where it stayed until 2012, the final year of this competition. Also important was that each year's data was broken up into three groups, one each for training, validation, and testing. In our discussion of machine learning for speech recognition (chapter 6), I emphasize the importance of ensuring there is no overlap between the data used to train the system

Figure 5.7: A labeled image from the PASCAL Visual Object Classes Challenge

and that used to test it. Careful researchers also use a separate corpus called the *validation* set, which is typically about the same size as the test set (and smaller than the training set). It is used when developing your program and helps ensure that you have not tailored the program to the testing data, which you are to use only at the end of the debugging process.

By 2012, the data set baton had been passed to a larger image set, the ImageNet Large Scale Visual Recognition Challenge (ILSVRC), this one organized by Fei-Fei Li, then at Princeton, now at Stanford [87]. Its first data set was released in 2010. It explicitly situates itself as a continuation of PASCAL VOC—the scene and labeling in figure 5.7 are from the PASCAL VOC but would equally well work to illustrate the ILSVRC. The latter, however, is much larger in terms of number of images (20,000 vs. 1,460,000) and number of categories (20 vs. 1,000). When you have 1,000 categories, you can cover the usual mammals (cats and dogs) and start working on, say, carbohydrates (french fries and mashed potatoes). This data set would prove its importance quite quickly.

5.6   Deformable Parts Models

I'm taking this break to talk about visual object challenges because the next object-recognition model dominated the field for about five years due to its successes in the PASCAL VOC from 2007 to 2011. This is the *deformable parts* model, which was created by Pedro Felzenswalb at the University of Chicago (now at Brown University) [27] and then perfected by Ross Girshick of Stanford [29].

The deformable parts model is an extension of the HOG model we discussed in section 5.4. This is best seen in figure 5.6, where we used the left-hand panel to illustrate HOG. The right two panels illustrate the deformable parts model (DPM).

In a DPM, we use standard HOG features as pointers to more complicated models that break the section of the object image into parts. In figure 5.6, we break a HOG feature for a person into eight parts, each of which can be thought of as a HOG feature itself. To do this, we compare the HOG model for, say, part 6 (it seems to be left shoulder as we face the image), against all positions in the bounding box for this person. Also, for each part we have a standard position for where this part goes and a scoring mechanism for how much we penalize a part for being in an unlikely portion of the image—say, if we put box 6 into the lower left. The score for "person" would then be the sum of the scores for its eight pieces, and the score for a piece is its HOG score plus its position score. In figure 5.6, the right-hand panel shows the positional scores for each part—the darker the area in the part, the more likely the center of the part is to be at that position.

When we learn a HOG feature for a person, a bicycle, or any other object, we learn the proportion of line orientations for lines in the object's bounding box. If we had labels where each piece of an object went, we could do the same for pieces. Unfortunately, our data set (PASCAL VOC) does not have such annotation, and we are forced to employ more indirect methods. Let us assume we are given the number of parts, the size of the bounding box for each part, and an approximate location. (This is not exactly correct, but it will give you the flavor.) We now go through the training data and assume the locations are correct. With the locations, we can compute the HOG statistics. Once we have the HOG statistics, we can drop the assumption that the locations are exactly correct and find the locations that, given the just-collected HOG statistics, best support the label and bounding boxes that *are* marked, thus improving the location information. We then go back and forth like this a few times, successively improving HOG statistics, then location statistics, and so forth.

HOG features are a good place to end this chapter because they are the high point of classical, pre–deep learning, AI vision research. In chapter 8, we discuss deep learning and show how deep learning revolutionized performance on the PASCAL dataset. The best performance numbers for pre–deep learning performance all use HOG.

6

Speech Recognition (1971–1985)

6.1 From Sound to Speech

Vision is the most important sense for us humans, so it is natural that it was the first to receive serious attention when AI considered computer sensors beyond keyboards. As for our four other senses, taste does not seem relevant, and smell only slightly more so. Touch is quite important, and our relative inability to deal with touch is a major impediment to progress in robotics. However, the problems involved with the sense of touch are sufficiently distant from the rest of AI, and from the material I know best, that I have excluded it from my history.

Our ability to hear is no doubt humans' second most informative sense, particularly because it encompasses our ability to understand spoken language—*speech recognition*. (The ability to make sense of the immediate world around us by interpreting the sounds it makes is called *audio scene recognition* and is normally treated separately from speech recognition. Again I omit it from our history.) Actually, most AI courses omit speech recognition entirely. For example, the text I mentioned earlier as my guide to what constitutes AI these days, *Artificial Intelligence, a Modern Approach* [88], by Russell and Norvig, gives it less than a page (out of about 1,000). (This is not a criticism. I have concerns here that are not present in an AI textbook—to show how many techniques critical to modern AI came to the field from speech recognition systems.)

Actually, while I am talking about that text, I should recommend the book's chapter endings to people wanting a more detailed account of all AI's intellectual currents. They link a huge fraction of AI's history

69

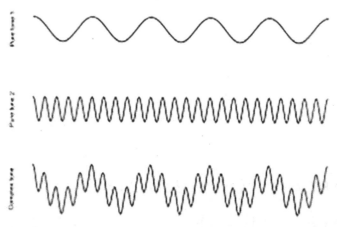

Figure 6.1: Sine waves combining to produce a more complex wave form

to the people and papers that make it up, something the idiosyncratic version presented here does not do.

Returning to sound, it is, of course, our brain's interpretation of *sound waves*—periodic high- and low-pressure waves in the air around us. These waves are, in general, very complex, and our auditory system starts the process of making sense of them by interpreting some of the complex waves as the conjunction of simultaneously resonating *sine waves*. Figure 6.1 shows two pure sine pressure waves and what would present itself to the ear when we play them simultaneously. The waves are called sine waves because they correspond to the mathematical definition of the *sine* function. If x is measured in degrees, then the top row starts at $90°$, when $\sin(x)$ has its maximum ($\sin(90°) = 1$). When $x = 270°$, sine reaches its minimum at -1 and then increases to its maximum again. The second row plots $\sin(6x)$, which cycles around six times as fast. The third has both sounding at the same time. In this case, a person would hear two pure tones, as if made by two tuning forks. If the higher sine wave was exactly two times the rate of the lower, the pitches would be one octave apart. Since these two tones differ by a factor of six, the two tones would be separated by more than two octaves.

Speech waves are considerably more complex. Figure 6.2 shows the signal the ear receives when someone says, "Eat your raisins outdoors on the porch steps." Despite this level of complexity, there is a well-known mathematical technique for breaking this signal into its sine wave components called Fourier analysis, after the mathematician Joseph Fourier. In speech recognition, it is standard to apply Fourier analysis to the signal and then represent the pressure signal as a vector with the intensities

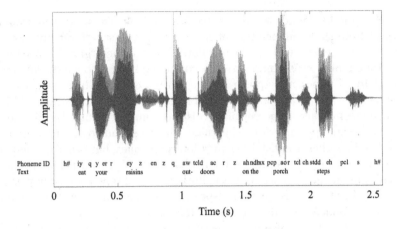

Figure 6.2: A speech waveform

of the pressure at a large number of frequencies. In the rest of our coverage, we assume this has been done, and the input to the rest of the speech recognition process is one of these vectors for every 1/100 seconds of sound.

Figure 6.2 also gives the *phonemic transcription*. A *phoneme* is a "perceptually distinct" unit of sound in a language. (I will return to the significance of "perceptually distinct" in a moment.) Various schemes exist for writing them down. You can think of the phonemes as a halfway point between the sounds and the words. The idea would be that if we can get a good phonemic transcription, you could then look up the sound patterns for each word and get the written words. There are about forty-four phonemes in English—there is no complete agreement on this number.

Unfortunately, speech recognition is not nearly that easy. First, word boundaries are not marked in the sound waves. People always complain that native speakers of a language are difficult to understand because they constantly slur their words together. This is true, because *everyone* slurs words together. Look again at the sound wave in figure 6.2. There is no separation between the "r" at the end of "your" and that at the beginning of "raisins." Or, then again, look at the transcription of "on the": it is one continuous sound. Contrariwise, the way we produce sounds leads to silences in the middle of words. Say the words "out" or "steps" from our sentence. To make a "t" sound, the tip of your tongue raises to the top of your mouth for a brief period, completely blocking the air, whether or not you are between words. During this period, naturally, no sound is forthcoming.

Nor do the sounds we make necessarily look like what you would expect from the pronunciation of the words. If you say "I left my dloves in the car," people will not hear "dloves," but rather "gloves," because (a) that makes sense, and (b) the combination of phonemes "d" followed by "l" does not appear in English, so our ears have learned to reinterpret "dl" as a combination that is allowed. Also, earlier I said that phonemes are a "perceptually distinct" unit of sound. By distinct, we mean that speakers of the language use the sound to distinguish one word and another. English uses forty-four phonemes, but other languages have different numbers and different phonemes. If you mispronounce an English phoneme, listeners find an alternative, and that is what they hear instead. In effect, this gives the speaker the freedom to be less careful when pronouncing words, and we all take advantage of this.

So, transcription of sound to written words turns out to be another problem that is much harder than it first appears. I remember early in my graduate years at MIT (about 1971), DARPA sponsored four groups to work on speech "understanding." DARPA is the Defense Advanced Research Project Agency—a branch of the US Department of Defense that has sponsored a lot of unclassified computer science research and engineering, including, most famously, the internet.

The researchers involved in this project were well aware of the difficulties I have been laying out here, and their idea was that people must use information about the topic under discussion to make up for the difficulties presented in identifying words by the sounds alone. Thus, the distinction between speech "recognition," which concentrates on the sounds, and speech "understanding," which gives equal billing to the use of the speech. I remember reading an evaluation of the project a few years later. It did not address this issue specifically, but contrary to expectations, the projects that came out looking the best were those that where more "sound based," with that by Bolt Beranek and Newman (BBN) coming out on top.

Actually, I have a good BBN story. BBN was a research and development company, since bought out by Raytheon, a large defense contractor, and renamed Raytheon BBN. Importantly, BBN was a main contractor for DARPA's internet project. The story starts out in 2002, but I must first introduce Scott Fahlman. Scott Fahlman is an AI researcher at CMU best known for his early work on knowledge representation. Actually, what I just said is false. Really, Scott Fahlman is best known as the inventor of the "smiley" :), these days also known as the "emoticon." He sent an email that used it back in 1982, and on the twentieth anniversary of this email, CMU publicized the event. Scott and I overlapped as grad students at MIT, and I mentioned to one of my grad students that "*I* know the inventor of the smiley," to which my student responded,

"*I* know the inventor of the at sign." Since she had worked at BBN before coming to Brown, I knew she was referring to the internet @, and since @ is more used than :), I was clearly one-upped.

As I was saying, BBN's speech system came out looking the best, but nobody really looked good. The CMU system restricted its speech input to commands about playing a game of chess, thus allowing the program to reason backward from the possible legal moves (or, even better, likely moves) to what was said. The standard joke, which perhaps was even true, was that if you walked into the acoustic room with the chessboard and coughed, the program would move pawn to king five (the most common opening chess move).

6.2 Speech Recognition and the Noisy Channel Model

All of this became history, because just as the DARPA program was winding down, leaders of two of the projects, James and Janet Baker, graduate students at CMU [1], and Fred Jelinek at IBM [43], started modern speech recognition with two innovations: the *noisy channel model* for speech, and the use of *hidden Markov models* (HMMs).

First, a bit of new math. The problem of speech recognition is to find the string of words that the speaker intended when they produced the sound waves that the listener heard. Or, expressed in terms of probabilities, the most probable string of words W given the acoustic signal A. Here, W is a sequence of words. Perhaps "I," "fed," "the," "cat." The speech signal A starts out as a continuous sound wave, as illustrated in figure 6.2, but as mentioned earlier, we break it up into 1/100-second pieces and represent each piece by a few numbers—the frequencies of the sine waves that make up the sound at that precise time.

The mathematical expression for "the most likely words given the acoustic signal" is

$$\arg\max_{W} P(W \mid A).$$

Here "arg max$_W$ *exp*" says, give me the value of W for which *exp* is at its maximum—in this case, give me the most likely words when A is spoken. So, this is the goal of speech recognition. Now, I don't know about you, but I am used to assuming that when I write down a mathematical symbol, I know how to carry it out, for instance, I can write down 5^3, and I know how to compute it: $5^3 = 5 \times 5 \times 5 = 125$. This is not always the case, however. The obvious way to compute the arg max is to go through all possible sequences of words and somehow compute the probability that the sequence will sound like A. So far in our discussion, we have not figured out how to do this. Indeed, the whole research process is aimed at this goal. All we have done is given ourselves a concise written form expressing what it is we want do.

We can now replace $P(W \mid A)$ using Bayes' law:

$$\arg\max_{W} P(W \mid A) = \arg\max_{W} \frac{P(W)P(A \mid W)}{P(A)}$$
$$= \arg\max_{W} P(W)P(A \mid W).$$

In the first line, we replace the conditional probability with the right-hand side of Bayes' law, and in the second, we omit the denominator. We can do this because all it does is *normalize* the terms on the right-hand side so that they sum to 1—so they are a probability distribution. But we don't need a probability distribution; we just want to know the W for which the expression on the right is the highest, and dividing them all by $P(A)$ does not change that.

OK, now look at this last expression slowly. We have divided the speech recognition problem into two pieces. This particular division of the problem is called the *noisy channel model*. The first part, $P(W)$, is called a *language model*; the second, $P(A \mid W)$, is the *channel model*. The way these ideas come together is that the second part "proposes" possible word strings based on the acoustic signal (the sounds), and the first part sees if these strings "make sense."

6.3   Language Models and Statistics

The purpose of a language model is to remove from consideration word sequences that might correspond to the sounds but make no sense. The classic problem is *homophones*—a group of words with different spellings and different meanings, but that sound exactly the same. Some standard homophones in English are "four" and "for" and "fore," or "our" and "hour." There is no way to distinguish "an our ago" and "an hour ago" based on sound. Marvin Minsky once told me that a relative of his visited him after also going to New York. Early on, his nephew asked him what a "nominal egg" was. It seemed that everything in New York cost a nominal egg. Minsky presented this with a straight face. I am not sure how long it was before I decided that I could not necessarily believe everything my thesis advisor told me. It also shows that language models can help distinguish between reality and tall tales. ("A nominal egg" is really "an arm and a leg" with a New Yawk accent.)

The reason the noisy channel model is such a good idea is (a) if you can construct a reasonably good language model, it will really help, and

(b) constructing a "reasonably good" language model is not as hard as it might first appear.

Despite my claim to the contrary, you would be in good company if you did not believe that language models are not that hard to construct. But, if you are willing to use a very bad one, one that assigns very low probabilities to strings of English words, it is simple—just assign the same probability to every word, and multiply together the probabilities of each word you encounter in the text.

Suppose English has a vocabulary of 50,000 words. We assign the probability of seeing the word "the" to be $\frac{1}{50000}$. We do similarly for "quick," "fox," and all the other words. Thus, the "probability" of "The quick" is $\frac{1}{50000} \times \frac{1}{50000} = \frac{1}{50000^2} = 50000^{-2}$. This gives us a probability of all two-word combinations of English words—they are all greater than or equal to zero, and they sum to one.

Actually, we have two problems here. We asked for a probability distribution over *all* English strings. The previous process gives a distribution for strings of length two. There is a different distribution for strings of length 3, 4, ..., and so on. We asked for a distribution over all English strings, not just strings of the same length. We'll solve this problem in a minute.

The second is that it is not true that English has just 50,000 words. For our purpose, we take as a "word" any string of characters in an English text between two white spaces, where we count a line feed as a white space. So if we saw, "Providence had a population of 189,692 in 2021," then "2021" would be a word, as would "189,692." With this definition of "word," there is no principled limit to the number of English words we find in texts, or, to use math-speak, the number of words is "unbounded." For the moment, we solve this problem by introducing a special new "unknown" word, "⟨unk⟩." Then, in any text for which we want a probability, after we have seen the first 49,999 words, any new word is replaced by this special word, and we have enforced our 50,000-word vocabulary limit.

We can make our language model look a bit less silly by instead giving each word (including the unknown word) its own probability. Take a large collection of texts of perhaps a million words, replace all excess vocabulary with ⟨unk⟩, then for each word compute, for example

$$P(\text{cat}) = \frac{|\text{cat}|}{1000000}.$$

Here $|\text{cat}|$ represents how many times the word "cat" appears in our "corpus." (When we use a body of text to gather statistics like this, it is traditional to use the word "corpus" for the text. "Corpus" is the Latin for "body.")

You may remember in our discussion of learning for computer vision, we said that it was important to divide our research data, such as sets of labeled images, into two or three subsets, one each for training, validation, and testing. The "corpus" described previously from which we estimate the maximum-likelihood probabilities is the training set, or training corpus. The text we are estimating the probability of would be the test set, or possibly the validation set.

So, if "cat" appears, say, twenty times in our corpus, we give it the probability $2 \cdot 10^{-5}$—we see "cat" about once every 50,000 words. This way of estimating probabilities—counting how often something happens and dividing by the number of "happenings"—is called the *maximum likelihood estimate*. At this point, we have moved into the domain of "statistics"—the cousin of probability theory that deals with estimating probability distributions.

Maximum likelihood estimates get their name because they have a very desirable property: they maximize the probability of the corpus from which they are computed—the training set. Or, to put this more precisely, of any individual word probability distribution, the maximum likelihood distribution for words assigns the training data the highest value. To compute it, we multiply the probability of the individual words together to get the probability of the entire training set. First note that the number you get is part of a probability distribution—if you summed the probability over all possible sequences of words, you get 1, and the numbers are all nonnegative. Next, to get some intuition for why this is the maximum, suppose you made, say, some word have higher probability by taking some of the counts and moving them to that word. If the word "kettle" appears once in the corpus, it is given the probability 10^{-6}—it appears once in a million words. Now, you might reason that we could double its probability if, say, we "stole" a count from "the" and gave it to "kettle," which would double the probability of the latter.

But, of course, this would lower the probability of "the," not by much, but from, say, $\frac{20000}{1000000}$ to $\frac{19999}{1000000}$. But, when computing the probability of our one-million-word corpus, the probability of "the" appears 50,000 times in the product, and the small decrement to its probability is enough to decrease the probability of the whole more than the doubling of "kettle." Admittedly, maximizing the probability of the training set is not the be-all and end-all. What we really want is a probability distribution that increases the probability of the test set. But it is a start.

We have so far created two very bad language models. Next, we take a giant step from very bad to just bad. The key is an equation we wrote in our earlier introduction to probabilities, the chain rule. Here is what it looks like when we use it to break the probability of a four-word sequence

$(W_{1,4})$ into four separate probabilities:

$$P(W_{1,4}) = P(W_1)P(W_2 \mid W_1)P(W_3 \mid W_1, W_2)P(W_4 \mid W_1, W_2, W_3).$$

Here we are thinking about a writer writing down the first four words of their essay, say, "The quick fox ran." We are defining four events, one for the choice of each word, so $W_1 =$ The, $W_2 =$ quick, $W_3 =$ fox, $W_4 =$ ran. We can thus talk about the "prior" probability that the first word is "The" ($P(W_1)$). Next, we want the probability that the second word is "quick," given that the first word is "The": $P(W_2 \mid W_1)$. Then, we need the probability that the third word is "fox," given the first two words are "The quick," and so forth. This equation says that the probability of all four coming together as the first four words is the product of those four probabilities.

The chain rule can be applied for as many events (words) as we like, but the problem is estimating these probabilities. By the time we have a test sequence of, say, 200,000 words (book length), and we need to estimate, say, the probability of the 10,000th word given the previous 9,999, we are in trouble. We look in the training corpus for how often we see the 10,000-word sequence and divide it by how often we have seen the 9,999-word initial subsequence, and, of course, we see neither of them. So, the maximum likelihood estimate is zero over zero—undefined.

The next step in our estimation process is to simplify the factoring given by the chain rule. Here is a version for the first four words that requires only finding identical sequences of length 3, not 4:

$$P(W_{1,4}) \approx P(W_1)P(W_2 \mid W_1)P(W_3 \mid W_1, W_2)P(W_4 \mid W_2, W_3).$$

In this version, the probability of the first four words is *approximately* equal to the right-hand side of the equation, the reason being that the fourth word is conditioned only on the previous two. So, to estimate the probability of "The quick fox ran," the probability of "ran" is conditioned only on "quick" and "ran." We then continue like this. That is, each word is conditioned on the previous two words, not the entire previous sequence. This model is called a *trigram* language model because it requires finding in the training corpus only identical three-word sequences to estimate the probabilities we need.

However, even a *very* large training corpus would not have all plausible three-word combinations. Consider the sentence, "Einstein bought cellos." There is a good chance that while Einstein played violin, he never bought any cellos, and if he did, nobody wrote about it. Now admittedly, we are unlikely to ever need this probability in our language model because it is no more likely to come up in the future than it has in the past. However, there are a *lot* of unlikely word combinations, and some do come up. For example (and I promise I did not plant this), else-where in this book I use the three-word sequence (ignoring punctuation) "stronger language nutty." The major problem with estimating language-model probabilities by counting how often the sequence actually occurs in a corpus is that we assign zero probability to any subsequence that includes the rare combination. Once we do that, the entire sequence is assigned zero probability (zero times anything is zero), and we cannot compare variants of the same zero-probability sequence.

The standard solution to this problem is to *smooth* the probabilities. Rather than computing the conditional probability of a word given the two previous words by counting occurrences in the corpus, we do the following:

$$P(\text{cellos} \mid \text{Einstein bought}) = .8\hat{P}(\text{cellos} \mid \text{Einstein bought}) +$$
$$.15\hat{P}(\text{cellos} \mid \text{bought}) + .05\hat{P}(\text{cellos}).$$

Here I have used $\hat{P}$ to indicate that a probability is computed by using maximum likelihood estimates based on our training corpus. What I have done is "mix" three probability distributions. If the three-word combination never occurs in the corpus, the probability of our entire sequence is now nonzero because we are mixing in the probabilities of the two-word sequence and just the prior of "cellos." (If, for some reason, "cellos" did not make it as one of the 50,000 most common English words, we would have replaced it by a special unknown word symbol "⟨unk⟩" in both our training and test corpora. Then, when we collect counts, we ask, how often have we seen "Einstein bought ⟨unk⟩?" and it has seen combinations involving "⟨unk⟩" many times. (As is discussed in chapter 10, real language models these days do something more sophisticated.)

6.4   Hidden Markov Models

The Bakers and Jelinek made two major innovations to the study of speech recognition, of which the first was the noisy channel model—splitting the problem into a channel model and a language model. The other significant contribution was the introduction of *hidden Markov models* to the speech recognition problem. Hidden Markov models (or HMMs) were described by Leonard Baum and colleagues in 1966 [2]. Again, this was not done in the context of AI, but rather from the perspective of electrical engineering, more specifically, the subdiscipline of signal processing.

A major problem in speech is going from the wavelengths of the speech signal (figure 6.2) to the phoneme sequence (a good intermediate

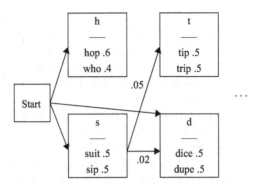

Figure 6.3: Representation of an HMM

point from which to suggest the words). An HMM for this task uses two sets of probabilities. The first is the probability of one phoneme following another. Suppose we have just head a "ss" sound, as in "stop." What is the probability that the next phoneme is a "t" sound, again, as in "stop"? There are about forty-four phonemes in English, so if any of forty of them were equally likely after "ss," the probability would be .025. But actually, "st" is a common combination, so this would be higher, say, .04. (Think of "store," "stove," "stamp," etc.) Contrast this with "d." Following "s," "d" is rare. Indeed, at first I could not think of any words with "sd" because I could not come up with words that started with "sd," and thinking of words with specific sounds in the middle is hard. (Fortunately, lots of Scrabble players use Google to search for words, and asking for words with "sd" together did provide a few. However, they are not common: "eavesdrop" and "misdirect" were the first two.) So, given an "s" sound, the probability of "d" might be .002.

The second set of probabilities captures the variability in the pronunciation of a single phoneme. Both "who" and "hop" start with the "h" phoneme, but the numeric representation of the sound waves are distinct. (Try saying both. I find it easiest to simply notice that the shape of my lips is different when I pronounce the "h" in each word.)

Figure 6.3 shows a graphical representation of an HMM. (Indeed, HMMs are considered one of the most basic of the so-called *graphical models*.) Of course, figure 6.3 is missing most of the forty-four phonemes, most of the arrows showing phoneme-to-phoneme transitions, and so forth.

With an HMM, it is not too hard to see how to compute the probability of a given sound sequence if you are also told the phoneme sequence: just multiply the probabilities of the phoneme sequence from one phoneme to the next using the first set of probabilities (on the arrows) times the probability of the phoneme pronunciation from the second. (I have indicated probabilities of a phoneme producing a sound vector with a word that uses that sound along with the probability.) For the probability of an entire sequence, you just follow the arrows from phoneme to phoneme, multiplying together all the probabilities you encounter.

But, of course, we want to go the other way: given the sound waves, what is the most likely phoneme string? First note that you cannot just follow the highest probability arrow at each point. Suppose we are at the "s" node, but the next sound is a "d." Several phoneme-to-phoneme probabilities are higher than "s" to "d" (the highest would be to a vowel, like "e"), but, of course, if the next sound is indeed a "d," those transitions will not result in the highest complete score because if we go to, say, the "e" node and consider the sounds available there, they look nothing like the "d" sound wave, which might even get a zero probability.

The reason HMMs are so popular is because, contrary to the way it was presented above, there is an efficient way to find the most probable sequence of phonemes given the sounds. It is called the *Viterbi algorithm*, after its inventor, Andrew Viterbi of the University of Southern California [105]. This was back in 1967, and, like the invention of HMMs itself, Viterbi's work was viewed as part of signal processing, not really AI. The key idea starts from the following point: If I know the phoneme that starts the most probable path to the end of the sound string, the most probable path *to* that phoneme is the correct start for the total path. Or, to put it another way, if at the ith point in the sequence we have the phoneme ρ_i (I am using p for probabilities, so I will use ρ for phonemes), you save a *back pointer* to that phoneme ρ_{i-1} that was one back in that path. If you do this at each step all the way to the end of the speech segment, you then follow the pointers back from the final step backward until you reach the "Start" state, and that is the total most probable path. You can then read off the phoneme state names in this path.

7

Learning Language (1985–2010)

This chapter is about research into (natural) language processing (NLP). More specifically, it is about the transition within NLP to a learning-based paradigm.

In what was another interesting twist in the AI story, speech recognition work led to the next big change in the foundational ideas of the field. In particular, in 1990, the speech recognition group at IBM, led by Jelinek, published a paper on *machine translation* (MT)—enabling computers to translate between languages [9]. This paper took the ideas used in speech recognition, most notably the noisy channel model and HMMs, and imported them to the MT problem. This chapter focuses on MT, as it has been a great AI success story. However, at the end of the chapter, we abruptly switch gears and look at syntactic parsing, one of many other aspects of language where probability, graphical models, and learning came to dominate the field.

7.1 Early Machine Translation

Machine translation has been a goal of AI researchers since the earliest days—the first attempts started in the 1950s. As with all the other AI challenges, the problem was much harder than people realized. The first things that people tried were variants of *word-for-word translation*—take each word from the source language sentence and replace it with its translation in the target language. Unfortunately, the translations produced where shockingly bad, as noted at the time by Ida Rhodes, an MT researcher at the National Institute of Standards [83]. A common joke at the time was that the saying "The spirit is willing, but the flesh is weak" (but expressed in Russian) was translated into English as "The vodka is good, but the meat is rotten."

To get some idea of how poor word-for-word translations are, if we translate "The spirit ..." (this time in English) into Basque (a language spoken the northeast region of Spain), we get

Espiritua prest dago baina haragia ahula da.

This is according to Google Translate, which uses the technology we discuss at the end of this book and is quite good. If you translate this back to English, you get

The spirit is ready, but the flesh is weak.

So, at least Google Translate thinks this is a good translation. But if you translate word for word, you get

Spirit perst there is bania meat weak yes.

(Here I used Google to translate each individual word separately.) It also translated the Basque for "flesh" into "meat," so perhaps the joke is not as far-fetched as I thought. And don't ask me about "perst" or "bania." According to Google, "perst" is indeed an English word, meaning "extremely cold" (when applied to a body part).

7.2   Sentence-Aligned Corpora

The IBM group had a radically different idea. Inspired by their speech recognition work, they proposed to use the following noisy channel model:

$$\arg \max_{E} P(E \mid F) = \arg \max_{E} P(E)P(F \mid E). \qquad (7.1)$$

This is the standard equation, but with the events E and F for English and French. This says that the most likely English E sentence given a particular French sentence F is the one that maximizes the product of the English language model times the channel model $P(F \mid E)$. This work seemed particularly audacious in that it further proposed that the computer *learn* to translate, by learning the needed probabilities from already existing material on the web, just as their speech recognition work learned the needed probabilities from human-aligned audio transcriptions.

Perhaps for this reason, their work did not attract many adherents, at least early on. (I remember a former IBM researcher who Brown hired away as describing the work as ridiculous, or perhaps he used the stronger language "nutty.") That the paper [9] was heavy on the math, and light on any intuition, probably did not help their case. Also, Jelinek, the leader of the group, had a reputation as a difficult person to get along with. He was famous for saying, "Whenever I fire a linguist, our system performance improves." I have avoided dealing in personalities in the book, but Jelinek's was so outsized, and the general view of him was so wrong, that I cannot resist talking about him. Years later, Fred and I became good friends. The quote is accurate—Fred had no internal censor. But at heart, he was a great, even gentle, person, but one with no idea of how things would sound when said bluntly in his gruff voice.

At any rate, while the ideas behind this paper did not take hold quickly, take hold they did, and a lot of them can be made intuitive. First, from their speech recognition work, they already knew how to construct reasonable English language models. The hard part is the channel model, $P(F \mid E)$. In this section, we look at their solution to a slightly easier problem—just learning translation probabilities for individual English words. In particular, we see how to learn $P(\text{chat} \mid \text{cat})$.

The IBM group assumed as input to their program a *parallel corpus*—a text and its translation into another language. We also assume the two texts are *sentence aligned*—we keep only those sentences where sentences in the two languages have a one-to-one correspondence. That is, a sentence in one language is completely expressed by one sentence in the other. This usually is the case, but the IBM group developed techniques for removing the occasional outlier. Given a parallel corpus of 100,000 sentences, it is amazing what can be learned.

The law in Canada requires that their parliament publish its proceedings in both of their official languages, French and English. Also, very early on, the government made the proceedings available online. Known as the Canadian Hansard, it is named after the (English) Hansard, which contains everything that is said on the floor of the House of Commons, and was first published by Thomas Hansard.

The IBM group first worked out an algorithm to sentence-align the French-English corpus. You look at the length of the first sentences in each corpus, and if they are approximately the same length, they are probably in one-to-one correspondence. If not, not. Standard algorithms track all the possibilities simultaneously.

7.3 Learning Translation Dictionaries

For MT, however, what you really need is word alignment. Assume each English word is the translation of one French word. To make things more

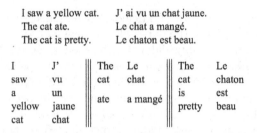

Figure 7.1: Sentence and word alignments

specific, assume the only sentences with either "chat" and "cat" in our sentence-aligned corpus are shown at the top of figure 7.1. Now figure 7.1 includes the word alignments, and with them, the problem is easy. To get the probability that "chat" is translated to "cat," we count how often "chat" is translated as "cat" (twice) and divide it by how often anything translates to "cat" (three times). (It is translated from "chaton," French for "kitten," once.) So, the probability is 2/3. This no doubt underestimates the probability that "chat" is translated as "cat"—it is more likely .9, or perhaps even higher—but I wanted to get across that chances are good that many words could translate to "cat." (Perhaps in French, we have named the cat, and "cat" is translated from a proper name.) Given a word-for-word alignment of a large translation corpus, we can thus get good ballpark estimates for all of the words. These are the maximum likelihood estimates that we discussed earlier in our brief discussion of statistics. Presumably, we would get a word-aligned French-English corpus by hiring a person fluent in both languages. We could then get the probabilities by doing both counts and dividing one by the other.

However, the IBM folks did something more clever: they created a program that learned the word-aligned corpus directly from the sentence-aligned version. To see how this can be done, suppose the person creating the word alignment was not fluent in French but, rather, was a native English speaker with two weeks of college French. (This is all, of course, a thought experiment.) To accommodate the translator's lack of knowledge, we allow them to align the words with probabilities attached when they are not certain. For example, in the first sentence, perhaps they are thrown off by the fact that "yellow cat" is not in the same order in French, so they mark that "cat" is aligned with "chat" with probability 0.9, and 0.1 with "jaune."

In computing our probabilities, we could elect to throw out any alignments that are not certain, but this might be a lot. We could also round the probabilities to either 0 or 1, but, in fact, the most logical thing is to take them as they are. We can still get the probability that "cat" is translated from "chat" by summing how often it is translated from "chat" (now it is 1.9 times) and dividing by the number of times it is translated from anything. (That is still three times, where 0.1 times it is translated from "jaune.") That would give us $1.9/3 = 0.63$. You can see that even now we might get plausible probabilities.

Now suppose our student knows no French at all. Rather they say, well, "cat" must be translated from *some* word, and the first sentence pair contains six French words. So, the probability in the second sentence that it translates from "chat" or "un" or whatever is $\frac{1}{6}$. We now do a similar calculation as last time—first, sum the probabilities that "cat" translates from "chat." It is $(\frac{1}{6} + \frac{1}{3} = \frac{1}{2})$. We then divide by the probabilities that cat translates from any French word. The latter is still 3, so the probability is $\frac{1}{6}$. This is very low, but still, contrast this with the probability that "cat" is translated from "mange." We are going to get a count of $\frac{1}{3}$ from the second sentence. But assume, as is reasonable, we have many sentences about eating, and, say, nineteen more use "mange." With these numbers, the count of "mange" to "cat" is $\frac{1}{3}$ and "mange" to anything is twenty, so the maximum likelihood probability of "cat" given "mange" is one in sixty. So, even starting with the assumption that anything could be translated to anything with equal probability, we are already seeing quite skewed results.

Furthermore, and this is what seems like magic the first time you try it, let's adopt the new probabilities and repeat the process. That is, we still assume that "mange" is translated from one of the three French words in sentence 2, but rather than apportion the probability equally, we do so in proportion to the probability we assigned for $P(\text{cat} \mid \text{Le})$, $P(\text{cat} \mid \text{chat})$, and $P(\text{cat} \mid \text{mange})$. The correct one, from "chat," is $\frac{1}{6}$, the incorrect "mange" is $\frac{1}{60}$, and since "Le" will have appeared in hundreds if not thousands of sentences, the conditional probability will be even smaller still. So, on the second iteration, the count apportioned to "cat" from "chat" is about .9, much higher than the .33 we gave it on the first iteration. Repeating this process a few more times, it will be close to $\frac{2}{3}$. which is about right for this slightly abnormal set of examples. The result of this process is a pretty good word-for-word translation dictionary.

We can compute yet further facts about French-to-English translation from a French-English sentence-aligned corpus. In fact, the IBM group designed five translation models, called IBM Models 1 through 5, each relying on more complex relationships in the data. And IBM was not alone in adopting this approach. By the late '90s, Google had hired Franz Och, who did his PhD work with Hermann Ney at Aachen University in Germany. Interestingly, Ney also started working on statistical MT after doing research on speech recognition.

7.4　Language Models Redux

Language models have the nice feature of being practically self-evaluating. We have introduced language models as a way of improving machine translation. But they are a small part of MT, and evaluating MT is difficult. (The best way is using them to translate the same set of sentences and simply asking a person which did the better job.) Fortunately, we can say that one language model is better than a second if, for a reasonably large chunk of text (or even better, lots of different chunks about different topics), one assigns a higher probability than the other. Since the texts we are using are presumably "good" in the sense that they written by people and, thus, do not have major mistakes (e.g., "cat yellow"), the higher probability it assigns to good English, the lower to assigns to bad, and, therefore, it is better at rejecting MT mistakes.

However, this works only if the text you use for testing has no overlap with that on which the language model was trained. In particular, if it does overlap, all three-word sequences in the testing data appeared in training, which virtually never happens in good (fresh) data. The result is the probability of the test data is mistakenly high. If we are using the probability of test data as a measure of language model quality, we get a result that falsely indicates the model is better than it really is.

Or, to put it another way, real data will have unseen combinations, even for trigrams. Earlier we saw how smoothing avoids zero probabilities in such cases, but it is inevitable that the probabilities produced by smoothing are still too low. Conversely, if the testing data appears in the training data, reporting such results can ruin your reputation.

Twice in my career, I have used contaminated data to test a language model. Fortunately, the first time my intuition told me not to believe the numbers I was seeing, and I just ignored them. (I had three years of sentences from the *Wall Street Journal*, but indeed, the second year had articles repeated from the first.) In the second case, I thought I had made a really important discovery on the creation of better language models and sent the results to some close colleagues. Fortunately, one of them had used the same data sets I was using and warned me that my test data could be contaminated with sentences from the training corpus. I checked, and sure enough, my "breakthrough" results were just due to bad test data. Since I had not published the results (emails don't count), my reputation for accuracy was bruised but not demolished. This is not to say that my papers are mistake free. I made a mistake in one that was caught by another researcher a few years later. The researcher wrote up a paper, and, wouldn't you know it, I was sent this second paper to review for publication! I told the journal editor that, indeed, my paper had a mistake and that this paper corrected it. However, I also pointed out that (a) the incorrect sentence was the only mistake in the paper—that is, nothing else in the paper depended on the mistake—and (b) it was not a very influential paper. (Google Scholar is a service that tracks how often a paper has been referenced by subsequent researchers in the area. As a rough measure, the higher the number, the more important the paper.) I then made no recommendation one way or the other on publication.

Here's another language-model personal story. In about 1988, I was invited to a DARPA workshop on language processing where I met Peter Brown, who had received his PhD at CMU and had worked with Geoff Hinton. At the time, he was working at IBM with Fred Jelinek. I remember the conversation for two reasons. I was still working within the symbolic AI paradigm, and Peter Brown started telling me about language models and how they were the right way to think about language. I was already beginning to have doubts about symbolic AI, so this conversation fell on fertile ground.

I also remember the conversation because Peter Brown joined a hedge fund named Renaissance Technologies a few years later when the CEO of Renaissance, James Simons, offered to double his salary. Peter Brown is now president of the company and presumably quite wealthy. About fifteen years ago, I invited Peter to give a talk at the annual meeting of the Association for Computational Linguistics—the grandparent of what are now many conferences specializing in AI language processing. I figured that his story might inspire others to follow his lead, and with enough hedge fund presidents, we would no longer need government funding. Peter thanked me for the invitation but explained that his talk would be a flop because everything he knew was either secret or boring!

7.5 Graphical Models and Grammar

All languages exhibit grammar—rules for how sentences are put together. I remember receiving explicit instruction on sentence structure when I was in grammar school (note the name!). First, we wrote a simple declarative sentence, subject-verb-object, like "Alice moved the tray." and were told to underline it. Then we drew a vertical line between the subject and the verb, followed by a half vertical line (stopping at the horizontal line we drew first) between the verb and direct object. See figure 7.2. If this was meant to improve my grammar, it was a total failure, but I suppose I learned the terminology. When my son was in fifth grade or so, I asked him if he had any lessons like this. He said that he had, and they also learned about prepositional phrases (as in "Alice moved the tray *to the floor*."), which they called "triangles." While I

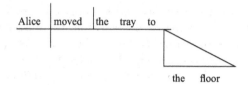

Figure 7.2: Diagramming a sentence in grammar school

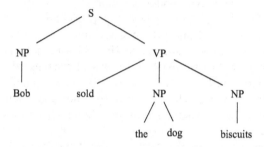

Figure 7.3: A grammatical tree

found grammar useless, I have spent much of my academic career trying to teach it to computers. The real purpose of grammar is not to improve your writing but, rather, to help construct the meaning of sentences from the meanings of their parts.

Almost every sentence you encounter is new to you. Take the last sentence of the previous paragraph. I am pretty confident you had no trouble understanding it, but even if somehow you have read a sentence with similar content, it is *very* likely that the exact sequence of words is novel. If they weren't, you might suspect plagiarism. The question this raises is how you understand novel strings of words. The answer must be that the pieces of sentences, the words, are not new, and that is sufficient to put together the meaning of the entire sentence. The thesis is that it is grammar that allows for this.

Consider the sentence "Bob sold the dog biscuits." The linguistic equivalent of underlining sentences and drawing triangles is the grammatical tree, as shown in figure 7.3. The entire sentence (S) is broken into a noun phrase (NP) and verb phrase (VP). Note that "sold" is a verb that accepts indirect objects, in this case, "the dog," so the verb phrase is broken up into the verb plus two noun phrases. It is the grammar of English that tells us this, and it also tells us that the first of the two has a semantic role of "recipient," not the direct object. (That is the second NP.) So this version of our sentence says that the dog bought the biscuits.

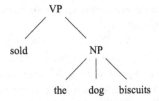

Figure 7.4: Tree for the second meaning of "sold the dog biscuits"

The reader may have noticed that I deliberately constructed this sentence so it is *ambiguous*—it has more than one meaning. Furthermore, the different meanings have different trees. The verb phrase for the second version has the form found in figure 7.4.

A grammar, then, is a set of rules for how sentences are built from words. For our sample, some of the rules we would need follow:

1. S → NP VP

2. VP → sold NP NP

3. VP → sold NP

4. NP → Bob

This form of grammar is called a *context-free grammar*, or *CFG*. The rules are to be interpreted as saying, for example, a verb phrase (VP) can be built as the verb "sold" followed by two noun phrases (NPs). The symbols in capital letters are called *nonterminals* because they have to be replaced using other rules to make a complete sentence. This is called a "context-free" grammar because all rules have exactly one nonterminal on their left-hand side, so the rule says that the nonterminal may be expanded using any of its rules *without regard to context*.

Now, remember what I said about the grammatical tree telling us how to put the sentence together from the words. Look at the word "biscuits" in the second version and how I have put it into the same noun phrase with "the dog." This tells us that we combine "the," "dog," and "biscuits" before adding it to "sold," as opposed to the version in figure 7.3, where we make "the dog" and "biscuits" into separate things and then add the two of them to "sold." Or, to put it another way, the first version talks about some particular dog; the second does not.

The idea is that if a machine can establish the syntactic structure of a sentence—if it can *parse* the sentence—that would tell us, at the very least, the order in which the meanings of the words should be combined. Forget trying to improve the computer's grammar (a *prescriptive* grammar)—we just need a grammar that captures the way the language is used (a *descriptive* grammar).

So, being able to parse garden-variety text has been a long-standing goal within AI natural-language processing. And, as we have often seen before, early ideas were imported into AI, this time from compiler construction, where people were interested in not parsing English but rather computer languages like FORTRAN. People found that context-free grammars were sufficient for most computer languages, and early on an efficient algorithm for parsing CFGs was independently discovered by three researchers in the area—the *CKY algorithm*—named for the three discoverers: John Cocke, Tadao Kasami, and Daniel Younger [49]. Starting with pairs of words, we find every possible structure permitted by our grammar for that substring. In each case, we look only for pair combinations, because once we have all of the smaller combinations, we do not need to look at them again when we create bigger pairs. (Take my word on this.) This is an instance of dynamic programming—storing the results for smaller subproblems because they can be reused when working on larger pieces.

This algorithm also finds *all* of the possible parses. These are expressed in the resulting data structure, so it is possible to tell if the sentence is *grammatically ambiguous*—there is more than one way to build it based on the grammatical rules.

Early on, natural-language processing researchers realized that while all the rules of English did not fit the context-free format, the ones that did not were comparatively obscure. Since efficient parsing with contex-free grammars is so easy, variants of CKY parsing were adopted for English.

Thus, at this point, the problem was simply to create a CFG that would assign intuitively plausible structures to the wide variety of English one encounters in books or newspapers. However, this task proved very difficult. Look again, say, at the first sentence in this paragraph. What is its structure? It is thirty-three words and punctuation marks long. By the time you get up to this length, figuring out the structure is hard, and the number of CFG rules needed to cover all the alternatives is huge. A lot of time was spent trying to find such sets, but everyone failed. The linguist Edward Sapir was famous for many important observations about language, not the least of which was "All grammars leak," by which he meant that it was not possible to write a complete grammar for a natural language.

Perhaps even worse, as you include more rules in the grammar, sentences that previously were unambiguous suddenly had extra structures that made little or no sense. For example, consider the phrase "five dollars a share." meaning that each share costs five dollars. The standard analysis is to create a new rule of grammar, NP $\to$ NP NP, where the first NP is "five dollars" and the second is "a share." Adding this rule is necessary for the sentence "The stockbroker paid five dollars a share," which is fine, but suddenly our "dog biscuits" sentence had not two structures but three. This new one has a noun phrase, "the dog biscuits," that itself is composed of two noun phrases, "the dog" and "biscuits." And longer sentences had even more—tens of thousands was commonplace. The field was a mess. Sentences either did not fit the grammar at all, or there were so many possible structures, little guidance existed for putting the pieces together, which was supposedly the purpose of the entire exercise.

The solution came from an unexpected direction. In 1994, a group at University of Pennsylvania led by Mitchell Marcus published the Penn treebank, a collection of one million words of text from the *Wall Street Journal* together with the structure of the sentences in the form of syntactic trees [61]. From the treebank, it is possible to read off all of the context-free rules necessary to assign the correct tree for all the sentences therein. This pretty much solved the grammar leakage problem.

Furthermore, you can count how many times each rule was used and then divide by how many times any rule for the grammatical category was used to get a probability that a given category would be expanded using a given rule. For example, if we divide how often NP $\to$ NP NP is used by the number of NPs in the corpus, we get the probability that any particular noun phrase is realized with the "five dollars a share" construction. This gives us a *probabilistic context-free grammar*.

We then solve the syntactic-ambiguity problem by requiring not *any* structure for a sentence but the most probable, which can be calculated by multiplying the probability of all the contex-free rules used in the tree. Furthermore, we have an efficient way to find this parse—the *inside-outside algorithm*. This gives us a grammar and algorithm that (a) have no leakage, at least in the sense they return *some* structure for any sentence you give it, and (b) return the single most probable parse according to the context-free grammar.

Unfortunately, returning the most probable parse is not the same as returning the correct parse. This is a good step in right direction, but, still, almost all the parses were wrong. Even if we look at an individual decision, the rule used to expand some particular nonterminal, the chance that the decision is correct is only about 75 percent, so by the time you have thirty or so decisions making up a tree, the probability that all are correct is very small: to a first approximation, sentences of eight words or fewer are mostly correct, all the others (90 percent) are wrong, with at least one mistake.

The final piece needed was *lexicalized context-free grammars*, as introduced by Michael Collins in 1996, also at the University of Pennsylvania [20]. The idea is to associate grammatical rules with particular words, which can be thought of as triggering the corresponding rules. For example, only certain verbs—sold, gave, take, and so on—can have indirect objects. You cannot say, for example, "Alice received Bill a gift." By giving a higher probability to rules when the sentence has the appropriate word (*lexical item* just means word) and lower when it does not, Collins raised the per-decision accuracy to 85 percent. And the next fifteen years saw the emergence of techniques that raised the accuracy to 97 percent, which is better than people can do at the task (only 95 percent). Assigning the absolutely correct structure to thirty-word sentence is quite hard, even for people. So, for all intents and purposes, the problem of syntactic parsing has been solved, at least for English. (Your author, together with his colleague Mark Johnson, was responsible for some of this improvement.)

8

Deep Learning (1989–2016)

When we last discussed perceptrons, Minsky and Papert had just pretty much squelched all research on them with *Perceptrons*, their volume on the mathematics and, critically, the limitations of this class of models. However, a group in San Diego centered around the mathematical psychologist David Rumelhart retained their interest in perceptron-like mechanisms and, in 1987, together with James McClelland of Stanford, published a two-volume set [86] titled *Parallel Distributed Processing: Explorations in the Microstructure of Cognition*, which was responsible for a resurgence of interest in the topic.

It also helped that Rumelhart, with Geoff Hinton and Ronald Williams, published a paper [85] that popularized *backpropagation* (backprop for short) as a method for training *multilayer perceptrons*, more commonly called *neural nets*. Backpropagation allows us to adjust the parameters of two or more layers of perceptrons connected together, one feeding the next. (You may remember the perceptron learning algorithm, but it worked for only a single perceptron or, equivalently, a single layer.)

If you have forgotten our brief introduction to perceptrons, it would be a good idea to review it—section 1.9. You should remember the final diagram showed a one-layer network that would classify the ten digits. Now, suppose we added a second layer of ten perceptrons just before the output, fed by the first layer, as shown in figure 8.1. To give you some idea of how training multiple layers can work, suppose we have an image of a 7, but the perceptron for "9" comes out highest. We go through all twenty perceptrons and modify their weights. The two layers on the right are meant to be *fully connected*—every perceptron in the first layer connects to every one in the second. In the diagram, I have shown the

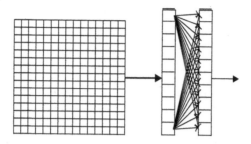

Figure 8.1: Multiple layers of perceptrons for ten-way classification of digits

connections for only the top-most and bottom-most because drawing all the connections would make it impossible to see what is going on. Let's say we are working on the bottom-most perceptron in the first of the two layers and considering the weight that connects it to pixel 14,14 in the center of the image. We can jiggle its value a little up or down, while holding every other weight steady, and see which direction causes perceptron 8 in the second layer to increase its value when compared to the others in that layer. We then increase or decrease our weight in the corresponding perceptron accordingly.

Now, this is not exactly how backpropagation works. We can use the calculus to figure out all at once whether to modify every weight up or down, and by how much, but the idea is the same.

I have left something out of the previous paragraph that is very important, but more complicated than I feel comfortable putting in a book aimed at a lay audience. It is not necessary for the rest of the book, so as I said once in our discussion of Bayes nets, if you want to skip the next paragraph, feel free.

You can show that if you do as I just said, *you will not get any improvement in the performance of the network.* For any two-layer fully connected network, a one-layer version exists that does exactly the same thing—so two fully connected layers give us no extra processing capabilities. The proof is not hard, but it is tedious, and to me, at least, gives no real "feel" for how it can be the case that (a) you get no improvement and (b) what I describe next fixes everything. The "fix" is easy: between the two fully connected layers, we put *relu* connections—rectified linear units. These connections produce a response like that shown in figure 8.2. Along the x-axis, we plot the value fed into the relu unit from the previous layer. The y-axis shows the value that is fed into the units of the next layer. The important thing is that the relu function is not *linear*. A linear function, something like $f(x) = 7x$, multiplies the input by a constant.

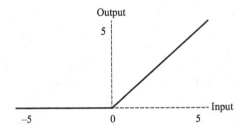

Figure 8.2: The response of a relu connection

In figure 8.2, if the line starting on the upper right continued straight at zero, the function would be $f(x) = x$, and that would be linear. So, whenever you see two fully connected layers, it is safe to assume there are relu connections between them. All these connections do is replace negative-value outputs with zero, but that is enough. We can use other nonlinear functions, but relu is very fast and, in most cases works as well, if not better, than the others.

Another, somewhat-on-the-easier-side set of images used to evaluate computer vision programs is the MNIST collection of 28×28 images of single digits provided by the US National Institute of Standards (NIST). The image of an 8 in figure 1.7 is from their collection. When trained and tested on their data, a single-layer ten-perceptron model classifies about 91 percent of the test images correctly. A two-layer model (with relu, from the last paragraph) gets about 97 percent correct. (If you remove the relu layer, the two-layer model reverts to the one-layer performance, as predicted by theory.)

8.1 Deep Learning and Computer Vision

The next big step toward the dominance of neural net (NN) technology— or *deep learning*, as it would soon be called—was the unbelievably pre-scient work of Yan LeCun of New York University, who, in 1989, while at Bell Labs, proposed using convolutional neural networks together with backpropagation to solve image-recognition tasks [53].

The reader may remember from section 5.2 that, in convolution, we repeatedly "apply" a small patch of numbers, called a kernel, to an image by multiplying each number in the patch by the pixel value at its corresponding location. The kernels can be designed to respond (assign high values) to different situations, most notably lines (intensity gradients). The lines could then be fed into detectors for more complicated information, such as HOG features.

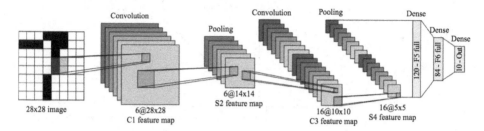

Figure 8.3: The LeNet neural-network architecture

The key idea of LeCun and his coauthors was to replace human-designed features by several layers of convolutional NNs and then learn their kernels with backpropagation. They named the NN architecture, shown in figure 8.3, "LeNet."

The image to be processed in figure 8.3 is on the left and the "28 × 28" caption underneath it indicates that the actual data is 28 × 28, not the 8 × 8 shown in the figure. The six squares to its right represent six different convolutional kernels, which (after backprop training) might be picking out lines at six orientations. This creates six 28 × 28 arrays, one each for the responses for one of the kernels. The third layer is labeled *pooling*, which we have not discussed yet.

If we step back and look at the overall flow of the network we can see that we start with a 28 × 28 image on the left and end with a 10 × 1 array on the right. This latter is just what we had in, say, figure 1.8 on the far right, where we had ten perceptrons, each computing the score that the image is one of the ten digits. So, the network can be thought of as starting with 28 × 28 values and reducing it to ten. Pooling layers are included in networks to step down the sizes on the way from 28 × 28 = 784 to 10. They do this by "pooling" information from several nearby locations in their input. Here, the layer is reducing a 4 × 4 patch in the first convolutional layer to the average value of the sixteen positions in question. The overall result is that our original 28 × 28 pixel image is now six 14 × 14 "images," each one picking out different features as dictated by the learned kernels.

The pooling layer is followed by a second convolutional layer. So far, we have considered using convolution only on images, and talked only about finding lines, but mathematically, the technique can be used on any array of numbers. LeCun and his colleagues correctly thought a second layer of convolution could pick up patterns in the lines found by the first layer. This second convolutional layer takes the 14 × 14 arrays and applies sixteen different kernels. It then pools again, and we get sixteen 5 × 5 arrays of numbers.

Initially, we think of the result of convolution as another sort of image, but the pixels in this image report not light intensity but line orientation (after one convolution), and then combinations of nearby lines (or perhaps texture) in the second application. However, once convolutions have been applied repeatedly, what is found at each point is sufficiently abstract that it then is convenient just to treat the information as a vector of numbers describing what is in the image. For example, one kernel might find vertical lines slightly tilted in the center of the image (the downstroke of the "7"). At this point, it is reasonable to switch to fully connected layers so we can get arbitrary combinations of features. Thus, the last two layers are fully connected. Finally, we have the output of ten perceptrons in the last layer, with the maximum value across the output units indicating the system's guess as to the digit in the image.

We train the model using backpropagation. This means increasing or decreasing each weight in each perceptron in the giant network so that the correct value at the end increases and the incorrect values decrease. This sounds like it could be time-consuming, and it is, indeed, the case that training a neural network takes much more computer power than using one. However, using the calculus, it is not necessary to laboriously trace the change at an early weight through all of the intermediate layers to find out how it affects the final output. We can work our way from the weights near the end, remember how they change things, and reuse the values. However there is no escaping that modifying each weight takes a few multiplications and additions, though since the additions are cheap compared to the multiplications, we just count the latter. If we assume we need ten multiplications per weight, the cost of a training step is $10n$ where n is the number of weights. So, typically, we measure the cost of training by this n.

What is n for LeNet? The answer is about 50,000. If that looks about right to you, you can skip to the next paragraph. Otherwise, here are some numbers. Pooling layers have no weights, since they don't have any perceptrons. Convolution requires only a few. For example, the first convolutional layer has six kernels. If each kernel is a 5×5 patch, that means the first layer requires $6 \times 5 \times 5 = 1,504$ weights. The second layer has sixteen kernels, so it would take another $16 \times 5 \times 5 = 400$ weights. However, the fully connected layers (also called *dense* connections) have many more. Consider the number of weights we need to go from the sixty-four perceptrons in the penultimate layer to the final ten perceptrons. Each of the final ten have sixty-four inputs each needing a weight, so there are 640 weights there. The other two dense connections are even more costly with about $120 \times 64 = 7,680$ to densely connect the 120 perceptron layer to the 64 layer, and about 40,000 for the remaining connections between the last 5×5 images and the 160 perceptron layer. As we see in chapter 10, the network is quite small by current standards (so-called large language models now are in the hundreds of billions). Nevertheless, given the computer hardware available in 1989, this was pushing things, and I would imagine the program was not setting any speed records.

Thus, already in 1989, we had a paper with a pretty much complete picture of how NNs could be applied to vision. But in one of the great quirks in the history of AI, *nobody seemed to notice.* I have mentioned that Google Scholar will tell you how many people have cited a paper. And over 11,000 have cited this one, which is large, though, given its importance, not as large as it could be. But that is not the interesting part. You can also get Google Scholar to break down the citations by year, and for the next twenty-three years, it received very few citations. Furthermore, as we mentioned in our discussion of computer vision in chapter 5, everyone else in the area in these twenty years was using SIFT and HOG features.

My analysis is that this is a lesson in the importance of standardized data sets, that is, at least one of the problems was that nobody else seemed to be using the MNIST digit dataset. People just did not read the article or, if they did, did not appreciate the results. You may remember that during this time, those researchers who would be most receptive of the LeNet results—those who were moving to the use of learning for vision recognition problems—were using the PASCAL VOC data set until 2011. Maybe they thought that what worked for digits would not transfer to more complicated images.

I don't think most nonscientists realize how hard it is to keep up with the scientific literature, which puts subtle but real pressure on scientists to discount others' work, since it means one less thing you need to read, or read carefully, which is nearly as bad. I definitely include myself here, and I kick myself for not cottoning on to deep learning earlier. I remember a conversation with my colleague Mark Johnson, who is quite astute, and both of us thinking that it was another fad. We had lived though previous incarnations of neural nets, and they had come to nothing, so we thought the new name, "deep learning," was a disguise to hide the fact that this was really just NNs again. Even after deep learning had shown its value in computer vision, I remember thinking that deep learning could be helpful in vision but only because there was nothing in vision that behaved like words do in language. Pixels are to vision as letters are to language. Words provide a good intermediate representation between letters and meaning. I thought that deep learning could provide something of the same thing for vision, but since we in language processing *had* words, we did not need deep learning. I was wrong.

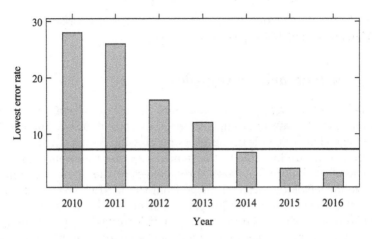

Figure 8.4: Error rate in the ImageNet Large Scale Visual Recognition Challenge by year

Back in computer vision, things did change in 2012 because of the results of 2012 ImageNet Large Scale Visual Recognition Challenge. Figure 8.4 shows the percent incorrect for the winners of the challenge from 2010 to 2016, so lower numbers are better. The heavy line at 7.5 percent is the human error rate of about 5 percent to 10 percent and shows that with 1,000 classes of objects, the task is far from easy. As noted in chapter 5, the first two years were still during the era when deformable parts models were popular. You can see the error rate is quite large compared to human performance and decreases only slightly from 2010 to 2011—a little over 1 percent. But in 2012, an over 10 percent reduction of error was achieved, down to 16 percent. This and all the subsequent competitions shown in the chart are with NNs. I am not sure how much we should trust the last few years, showing super-human performance. It may just be computer superiority at simply remembering all the 1,000 categories. But even matching our human ability in object recognition is a significant accomplishment.

The 2012 winning entry was by a group at the University of Toronto led by Geoff Hinton, one of the major figures in our history [52]. The lead author on the work was Alex Krizhevsky, then a graduate student. As has become the custom in neural net research, the organization of the network takes the first name of the lead author, so this program is called *AlexNet*. I do not show AlexNet, as it is pretty similar to LeNet but with more layers—two convolutional, two pooling, and three dense layers for LeNet, six convolutional, three pooling, and three dense for AlexNet. Furthermore, all subsequent improvements in ILSVRC scores were primarily the results of larger and larger networks. ResNet, the 2015 winner, had 154 layers in total.

8.2   Adversarial Examples

While deep learning programs dominate the area of visual-object recognition, they do have puzzling defects—they can be fooled. You can take an image that a deep learning program classifies correctly, make modifications that are invisible to the human eye, and the resulting image is misclassified. Figure 8.5 shows three images of the number 3. The one on the far left is the original image from the MNIST test set. It is correctly identified as a 3 by a program your author wrote. The one in the middle is an "adversarial example"—it is identified as an 8. The right-most image is an exaggerated version of the middle one, where the same changes that turn the left-most image into the middle one are made six times larger so they can be seen.

How is it possible that we can spoof the classifier so easily? Looking at the left-hand and right-hand versions, you can see that (a) whereas the original background is a uniform black, on the right there is a difference between pixels. Similarly, in the original 3, the writing of the digit, which is light, always shades from middle-high to maximum brightness. This is not the case on the right. The middle 3 moves very slightly toward the unusual qualities illustrated in the right-hand version. I say "unusual" as none of the examples in the training and test programs have these properties. Thus, the classifier has had no firsthand experience with such images and does not "know" how they are to be classified, making it easy to trick it. This phenomenon was discovered by Christian Szegedy of Google in 2014 [101].

We obtain these examples by first training our NN and then using it to classify an example. If it gets the example right, we then set about

Figure 8.5: An adversarial image of a 3

spoofing the program. Remember how backpropagation trains our network. It modifies each parameter in the NN. When holding all of the other parameters fixed, it changes the selected parameter so that the probability of the correct answer is increased. Well, now we hold *all* the parameters fixed and, for each pixel, determine if increasing its value (or decreasing it) changes the class assigned to the image. Having done this for each pixel, we get an array where at each point we specify which direction to tweak the pixel value—up or down. Then, we modify the pixel by some fixed value. If the brightness of a pixel varies between zero and one, we first try modifying by, say, .05 (5 percent). If this is not sufficient, we increase this to 0.1, then 0.15, and so forth. In fact, by the time we reach 0.25, all of the images have been thrown off.

Adversarial examples are of interest for two reasons. First, they say something interesting about our NN object-classification programs— they are clearly not working they same way people work since we are not thrown off by these machinations. Second, it could be argued that deep learning object-recognition programs are inherently unreliable because an adversary could cause them to spit out incorrect results. While I agree with the first of these, the second is, in my estimation, doubtful. The reason is simple: to carry out the deception, the adversary must get ahold of my images before I apply the classifier, and nobody has explained to me how this is to be done. More generally, if an adversary can waltz in and modify files on my machine, I have a more serious security problem than my images being misclassified.

Only one of the attacks mentioned in the literature actually makes any sense to me as a possibly serious security problem: when the adversary changes the world before I take my pictures. The standard example is modifying a stop sign so the NN thinks it is, say, a "no left turn" sign, thereby making any computer driving program quite dangerous. Some examples of signs that have been defaced to cause this confusion are shown in figure 8.6 [25]. But, the solution is obvious: drastically increase fines for defacing stop signs, and use the money to finance a rigorous program of stop sign clearing. OK, I am not sure if I am being facetious here, but the entire problem seems to me so conjectural that I find it hard to take it seriously.

8.3 Graphics Processing Units and AI Hardware

Deep NNs require billions of computations to train on a data set, so a laptop is not the computer of choice for this task. A standard laptop has one or more *central processing units*, or *CPUs*. Let's assume just one. It also has a bank of storage locations—the "memory." When you read

Figure 8.6: A stop sign that has been defaced to cause a misclassification

that a laptop has thirty-two gigabytes of memory, you are being told how much your laptop can store without going to any external device. The CPUs and memory make up the laptop's *hardware*. (We will ignore the laptop's input/output devices, like a keyboard, thumb drives speakers, and so on.) A key realization back in the 1940s was this memory could be used for two distinct purposes. The first was the instructions that turn our general-purpose laptop into, in effect, many special-purpose machines like a calendar program—the software. The other is the data on which these programs operate (the entries I have put into my calendar). The CPU cycles through the following three-stage process:

1. Retrieve a software instruction.

2. From the instruction, locate and retrieve the data on which to operate.

3. Carry out the instruction.

Early on it was realized that, contrary to what you might expect, part 3 is fast, and parts 1 and 2 are where the time goes. Or, to put it another way, once you had gathered the necessary information, processing it is fast. As a result, a lot of the transistors in a CPU are devoted to figuring out how to get data and instructions into and out of the CPU as fast as possible. We can do this in many ways, but the most common is to cache (temporarily store) the data on the CPU, where it can be accessed without the delay caused when the data has to be retrieved from main memory. However, doing this involves guessing what will be needed, which generally is hard, and when the cache does not have what the CPU wants, things essentially grind to a halt.

(I just used the word "transistor." Do people these days still know what a transistor *is*? In the old days, there were things called *transistor radios*. The first radios did not have transistors but, rather, things called *vacuum tubes*. These tubes were also inside computers. Vacuum tubes and transistors can act as switches, and it was a big deal when transistors replaced vacuum tubes in both radios and computers. The population at large did not notice the computer changeover, but the "transistor radio" could be battery powered and fit it your pocket! It may be possible to buy an individual transistor these days, but don't look inside your laptop or cell phone; they are there, but invisible, because they have been miniaturized so millions can fit on the chips.)

In the early days of the field, AI researchers used much the same computer hardware as everyone else. I have not mentioned it, but by the mid-1960s, almost all AI programming was done in *Lisp*. Lisp, which stands for *list processing*, is a programming language optimized for symbolic processing just as, say, the early programming language FORTRAN is designed for numeric processing. Lisp ran on what were vanilla computers for those days. Building a computer to handle larger programs amounted to putting more and more transistors on the chip. And, as we have just mentioned, a lot of those transistors were designed to speed moving information to and from the CPU.

(An interesting Lisp story: Lisp was sort of invented by John McCarthy. I say "sort of" because McCathy was working on mathematical problems in the theory of computation. This involved the creation of a "universal function." In computer science, something is called "universal" if, in principal, it can compute anything. To McCarthy, this was just a mathematical idea, but one of his students, Steve Russell, thought of programming this function on a department computer. McCarthy's reaction initially was somewhat negative, telling Russell that "he had theory and practice mixed up." These days, McCarthy is given the credit for inventing Lisp, but Russell is given a good second billing.)

Returning to the implications of how CPUs work, in convolution we multiply some pixel values times the convolution kernel—the small patch of numbers that specifies what this kernel is looking for, such as, a line of some sort. In convolution, we do these multiplications and additions over and over on equal-size patches from the scene. Furthermore, the additions and multiplications can be done in almost any order. Thus, if we have a group of processors that have access to a 100×100 region of a scene and, say, ten 5×5 kernels to apply to it, we have all the data needed to perform about $100 \times 100 \times 10 \times 25 \times 2 = 5 \times 10^6$ multiplications and additions without any further main memory access. Other than a convolution addition that has to be done after the local multiplication, all of the operations could be done at the same time. Remember, applying a convolutional kernel of, say, size 3×3 to a correspondingly sized patch of an image first requires nine multiplications of the kernel value times a pixel value, followed by an addition of the nine multiplication results. Thus, if we had, say, 1,000 processing units, we could speed things up by a factor of 1,000.

More generally, all of the major mathematical operations in deep learning are forms of *matrix operations*—mainly matrix multiplication and additions—which, in turn, involve a lot of plain old arithmetic. "Matrix" is just a fancy word for a two-dimensional "array" or table. We said that an image is a (two-dimensional) array of numbers. We can equally well say it is a matrix (the plural is "matrices"). Through a fantastic bit of luck, graphical computer games, or, for that matter, any computer application that requires computer graphics (manipulating images), also depends on matrix operations. So, any machine designed for computer gaming has a *graphics processing unit*, or *GPU*, besides its standard CPUs. You can think of a GPU as 1,000 slow CPUs—say, one-tenth the speed—but because there are so many of them, when they are set to a task for which they are designed—matrix operations are the most common—they are 100 times faster than CPUs.

At that time, about 2005, two very large GPU manufacturers were in business, Nvidia and AMX. In 2007, Nvidia released *CUDA*. You can think of CUDA as a programming language for programing 1,000 computers as if they were one computer. This allowed deep learning researchers to speed up their programs by factors of 100.

Since then, AI has become a booming business, with a multicompany race to build the next generation of still faster AI hardware. The basic idea is to make the hardware increasingly specialized to AI tasks. One we have seen is convolution. Another is a recent scheme called *transformers*, to be discussed in chapter 10. Since these processors are specialized for AI and not graphics, the new term for this kind of processing unit is *AI accelerator*. Accelerators typically have the equivalent of several GPU processors.

8.4   Distributed Representation of Words

Until now, we have looked at deep learning as applied to the interpretation of images. While this could be justified in that vision is our most important sense, sound, especially through speech and language, runs a respectable second. However, our emphasis on vision was not so much about its importance as the fact that deep learning first showed its prowess in that area. After all, visual information comes as light intensities, which are just numbers, and numbers are at the center of how perceptrons, neural networks, deep learning, and so on all work.

Language, however, is mostly symbolic. Indeed, words are the prototypical symbols. Consider the word "mirror." Until a minute ago, nothing about the word itself gave me any clue as to its meaning—it was a pure symbol for an object that reflects light coherently. (But a minute ago, I looked it up, and it derives from the Latin *mirari*, to admire.) So, how can we process words with NNs?

To make this more specific, consider creating a simple language model. We first associate a number with each word in our vocabulary. (Again, we propose that we fix our vocabulary at 50,000 words, so the numbers are from 1 to 50,000, with 1 being the number for aardvark, or however it works alphabetically.) Next, we create an array with 50,000 entries, and we give each word a *word embedding*.

A word embedding is a sequence of decimal numbers designed to represent a word within a neural network. The number of decimal numbers in the list for a word is up to the programmer. We can imagine using twenty, though in current big neural network language models (LMs), 1,000 would be a standard choice. Initially, the word embeddings are random numbers with no particular relation between a word and its embedding. As we will see shortly, it is only when we train the LM that the embedding somehow reflects its word. The 50,000-length array we just mentioned stores the embeddings, so the array is $50,000 \times 20$. We go through our LM corpus looking up each word and, in effect, replacing it with its embedding.

Let us assume we are using a *trigram*—literally "three words"—language model, as described in chapter 6. Remember that trigram is the technical term for three-word combinations, which come up when we condition the probability of a word on the two previous words. (Once again, we can estimate the probability of a word given the two previous words by counting how often the three words occur in sequence and dividing by the number of times we encounter the first two of those words, whether or not the third appears as well.)

The neural net would look something like that shown in figure 8.7. On the left are forty inputs, twenty decimal numbers each for the first two words of the trigram. So, if the first two words of our text were "The yellow," the inputs would be the twenty numbers representing "The" and the twenty for "yellow." We feed these numbers into some number of perceptrons (here just forty) and do this some number of times (here just once). At the end, we have a bank of 50,000 perceptrons, corresponding to the 50,000 words that can be the next one in the text. Each of these 50,000 words produce a number. During training, we adjust

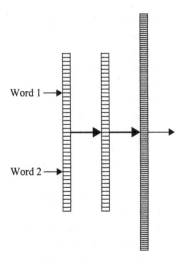

Figure 8.7: A neural net for a trigram language model

all of the perceptron parameters to make the one corresponding to the
true next word come out slightly higher and all of the others come out
lower. Now, when we say "all the parameters," we definitely mean to
include the numbers in our word embeddings. Again, to keep things simple, assume that the forty inputs connect directly to the bank of 50,000
output perceptrons—the middle layer in figure 8.7 is removed. And suppose the next word after "The yellow" is "aardvark" (which we presume
is in position 1 in the right-hand layer). We are going to change the
numbers in the embeddings for "The" and "yellow" so that, holding the
rest of the network constant, the output of perceptron 1 increases. We
do the same thing for all the other perceptron parameters as well, but
here I am concentrating on the embedding parameters, because when
we make these modifications, something very interesting happens—*the
embeddings of words that have similar meanings become more similar.*

To understand how this can be, first, what do we mean by two
sequences of numbers being more or less similar? The answer is: treat
each number as a position in space, and the similarity of two sequences
is the distance between the two positions.

OK, suppose we have sequences of length two: [3,6] and [4,3]. We
compute the *Euclidian distance* between two points as follows:

$$\sqrt{(3-4)^2 + (6-3)^2} = \sqrt{(1+9)} \approx 3.16.$$

So, to make two embeddings more similar, we simply try to make the
first numbers in each closer to each other and the second numbers closer

to each other. If we have length twenty embeddings, we want all twenty pairs closer.

Now, consider what is going to happen for, say, the embeddings of "automobile" and "car." Because these two words are, in fact, very similar in meaning (they are a classic example of synonyms), the words that they follow, or that follow them in sentences, tend to be the same. They sometimes follow "red," almost never follow "hydrant," and often follow "the," but rarely precede it. If you think about it, every time two words precede or follow the same third word, their embeddings should become more similar because we want the values for the following word to increase. There is a famous saying in linguistics, "You shall know a word by the company it keeps," so similar words have similar neighbors [108]. That this is the case was really brought home in a paper by Tomas Mikolov in 2013 [69]. Mikolov used a somewhat more complicated method for finding embeddings, but again he depended on the similarity of meaning leading to the similarity of neighboring words.

NN language models have the great property that we do not have to worry should sequences of words never appear in our training corpus. First, as mentioned, we have a standard way to turn scores for words into probabilities such that words with high scores have high probability, and while low-scoring words have low probabilities, they never go to zero. Also, to continue with the example of "car" and "automobile," suppose we had never seen "He buys used automobiles" but we have seen "He buys used cars." If, as we said, "cars" and "automobiles" have nearly identical embeddings, they will also have similar scores and, thus, similar probabilities.

Not only do embeddings cluster according to meaning, but they often behave understandably when they are added and subtracted. In what I found absolutely jaw-dropping, Mikolov took his embedding for "king," subtracted the embedding for the word "man," and then added the word "woman." Now compare the resulting number sequence to all 50,000 embeddings. Believe it or not, the result is closest to "queen." Or, again, take "Moscow," subtract "Russia," and add "China." The nearest embedding is for "Bejiing." So, we have reduced word analogy problems to simple vector arithmetic.

Before moving on, we need to draw out one very important implication in this treatment of words. I have expressed my unease with the identification of concepts with symbols, starting in chapter 2. Words are not the same thing as concepts, but they are close. Word *meanings* are concepts. Homonyms are words with two or more different meanings— "bank" as a savings institution or part of a river, "watch" as a time piece or as a a verb, "to observe." If we could simply take what seems like a relatively small step from embeddings as our representation of words to

embeddings as the meanings of words, we have a new notion of what
concepts are that is completely opposed to, and more flexible than, the
physical symbol hypothesis that Newell and Simon enunciated and we
saw back in chapter 2. Furthermore, coming up in section 10.1, I argue
that a new way of handling embeddings has gone a significant distance
in this direction.

8.5   Recurrent Neural Nets

Neural networks are *directed graphs*, vertices that are connected by *directed* edges, which is to say the edges are not lines but arrows. In our neural
networks, the vertices represent computational units, and the edges show
the flow of information (numerical results) from one computation to the
next.

Until now, our NN graphs have been *acyclic*—not containing cycles.
A directed graph contains a cycle if, starting at some vertex and following arrows in the correct direction, it is possible to return to the
starting vertex. An example of a cyclic NN (also called a *recurrent* NN)
is shown in figure 8.8. A bit of study shows how this network can be

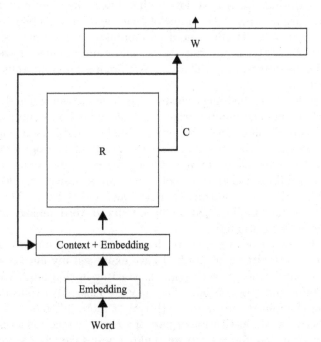

Figure 8.8: Language model using a recurrent NN

used as a language model, processing language one word at a time. The next word of a text enters at the bottom and is immediately replaced by its embedding. At the other end, at the top, the box labeled W is a layer of 50,000 perceptrons that compute the scores or probabilities for every word in our vocabulary. During training, the network outputs the highest-valued output to be compared with the actual word that follows.

In the middle, we have a cycle. The perceptrons in R output (to the right) a vector of numbers C that we have called the "context." Most directly, the context is the basis for the NN's guess at the following word. But you can see that it also cycles around to the point where it is combined with the next word for the subsequent prediction. As with other neural LMs, all the weights are trained with backpropagation.

Recall our first attempt at a NN language model, shown in figure 8.7. It was based on the trigram model—the probability of the next word was conditioned on only the two previous words. In contrast, the recurrent version, at least in principle, has information from indefinitely far back in the context vector, being passed on inside the context vector.

In practice, the first attempts at recurrent LMs did not work well. Information did not easily get into and out of the context vector, so the model quickly lost track of where the sentence had been. Things improved significantly in 1997, when Sepp Hochreiter and Jürgen Schmidhuber developed a technique called *long short-term memory* [38].

In essence, they noticed that in a basic recurrent network like the one shown in figure 8.8, the context vector C gets rebuilt at every iteration. Their technique instead has as the default that C gets passed from one iteration to the next without change, unless the incoming word, together with the recurrent array R, decides to add or remove sections of C. This practice drastically improved information retention in the networks. Recurrent networks with long short-term memory were, thus, the technique of choice for language models and, indeed, all tasks involving language, including machine translation. In 2016, Google started replacing its statistical MT system, which was based on the technology described in chapter 7, with neural network methods. (I remember chatter on the web because people noticed their translations improving from one week to the next.)

8.6 Autoencoders and Generative Models

We could go several different ways in our description of deep learning. I am going to pick *autoencoders*, because, as I write this, I am also working on chapter 10, which is on recent results, including the program DALL-E.

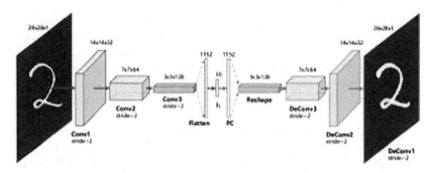

Figure 8.9: Schematic illustration of an autoencoder

To understand how DALL-E creates pictures when you give it an English request, it would be useful if the reader knew what an autoencoder is, so let's head over in that direction.

An autoencoder is a NN that learns a reduced representation of information (here, we consider pictures) by training a NN to reproduce the information, despite there being an information bottleneck between the input and output (see figure 8.9). On the left, we input an image, and much as if we were engaged in an image-recognition task, we put it through a few convolutional layers. You might go back and take a look at the diagram of LeNet. It takes the 28×28 image and creates a vector of length ten—the ten-way classification of the image content. If we concentrate on the left half of the autoencoder in figure 8.9, we see that it turns the 28×28 pixel image into a size-ten vector, but this version is then fed into a second network that reverses the process with the goal of just reproducing the input. However, the reproduction starts with much less information. Furthermore, if you focus on the two images, you see they are not identical, and the one on the right is more "generic." The point is, the small vector in the center is not sufficient to record the minor details of the "2" shape on the left, but it does record the most important points.

An autoencoder is almost an example of a *generative model*—a probabilistic model that can generate new data from a particular distribution. Imagine the collection of all 28×28 images of digits. The MNIST data set has about 70,000 such images and can be thought of as an approximation of the imaginary collection of all such images. Now, some of these images are common, a "1" as a straight line right down the middle; some are idiosyncratic and, thus, uncommon—a "1" as a nearly horizontal line. If I is an event of picking one of the images randomly, then $P(I)$ is the probability of picking a particular one, say, the 8 shown in figure 1.7. The idea is that if we give the hypothetical generative model a randomized input, it should return an image from the collection, with common images returned more often, like the vertical one, and horizontal ones hardly ever.

If you are willing to accept an awful generative model, creating such a thing is easy—have the model take $28 \times 28 = 784$ random pixel values, repackage them into a 28×28 array, and call that your output. Eventually, it will output all of the digit pictures, and there you go. Of course, most of the output will be just random noise, and the meaningful digits won't be output with the right probabilities, but I told you it would be awful.

Now, if you have trained an autoencoder, you can do somewhat better. After training, throw away the left-hand side of figure 8.9, so now the input is the small vector that was the "bottleneck" of original model. Rather than give the ridiculous model 784 random numbers as input, you give the half-autoencoder just a handful equal to the size of the bottleneck. We then run the remaining half of the autoencoder and return the 28×28 output as our digit image. Because it has been trained to output copies of good digits from the numbers it receives, a greater percentage of the random combinations should look reasonable (though it will likely still be a small percentage).

A still better generative model is the *variational autoencoder* [51] created by Diederik Kingma and Max Welling of the University of Amsterdam in 2014. This is what the DALL-E image generator I mentioned earlier uses. I am not going to explain it, except to say that it is further trained to produce reasonable variations on the image input given a random number.

9

Reinforcement Learning and the Game of Go (1990–2017)

9.1 Go

The success of the chess program Deep Blue in its 1997 match against Garry Kasparov was important because chess is the standard intellectual game in the West and, thus, a natural benchmark. However, in the East, the game of Go gets the most respect, and the success at chess did not transfer to Go or, in fact, to any other AI task. Go-playing programs were very weak in 1997, and nobody anticipated any change.

As noted earlier, Go is a two-player game of perfect information. It is also a board game, though crucially, the board has more positions, $19 \times 19 = 361$ compared to $8 \times 8 = 64$ for chess. Initially, the board is empty. The player playing black places a black pebble on an empty board location, the other uses white, and they alternate. Once placed, a piece does not move, except to be removed from the board. Black goes first, but white gets a small boost in the scoring to make up for black's advantage for going first. When a player completely surrounds some area, the player's score increases by the size of the surrounded area. This is called "capturing territory," and modulo the scoring preference given to white, the player who captures the most territory wins. A mid-game Go board is shown in figure 9.1. Go has been played in China for about 2,500 years—it makes Western civilization look very young indeed.

Go proved so impervious to anything else happening in AI that it was a surprise to everyone when the AI startup, DeepMind, created AlphaGo,

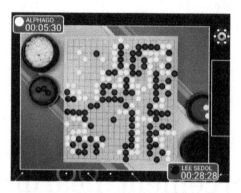

Figure 9.1: A Go board position from the AlphaGo vs. Lee Sedol match

the program that in 2016 beat Lee Sedol, a Korean professional, then ranked number three or four in the world. (Alpha is the first letter of the Greek alphabet and is frequently used in computer science to indicate the first version of, say, a program. Thus the name, AlphaGo.)

That the key to DeepMind's success was its use of deep learning [97] had a major impact on the field, your author most definitely included. After all, the initial impetus for the deep learning wave was not game playing but computer vision. When a technique helps in what appears to be a far distant area of the field, it suggests there is something "right" about it.

9.2   Neural Nets for Go

AlphaGo's key idea is to treat the Go board as a picture. Think of each board position as one pixel, which may have one of three values: empty, occupied by white, or occupied by black. Now rather than identifying a "picture" as one of, say, dog or cat, we label it as a board that is leading to a win for white or black. Given the record of a game, for each board position we know (a) the move made by a player (mostly good amateurs) in response and (b) which player won after starting from that position. Thus, we can train two neural networks—one for picking moves, the plausible move generator, and the other for static board evaluation, as discussed in section 4.1. So, the overall structure of AlphaGo is the same as that of other programs for two-person games of perfect information:

- suggest some plausible moves (or all legal moves),

- look ahead in the game tree some number of ply (move/counter move pairs) using alpha-beta search,

- when the search reaches the preset maximum search depth, the final game states are assigned values by the static board evaluator.

AlphaGo's neural networks are twelve layers deep, and it is virtually impossible to look at them and say much of anything about why AlphaGo makes the moves it does. The program, however, does have one easily understood point of style—it plays "conservatively," in that it often makes moves that reinforce its strong points in the game rather than moves (more likely to be chosen by a person) that could capture more territory. It makes sense that it does so, however, as it is trained to maximize its probability of winning rather than the amount by which it wins. It is trained using reinforcement learning, to which we turn next.

9.3 Reinforcement Learning

Reinforcement learning (RL) is a method for teaching a program how to achieve goals when it is given *rewards* for achieving certain desired states of affairs. We typically speak of the program as the *agent*. A canonical situation would be an agent that can go around a grid by moving either left, down, right, or up at each moment of time. If it arrives at certain "goal" states, it receives a reward. (Or, it can be "punished" if the "reward" is negative.)

An RL problem consists of the aforementioned reward function, a set of states (one of which is the *starting state*), a set of possible moves that take the agent from one state to another, and a *discount*. The discount is a number between 0 and 1, and the reward one receives is actually the product of the value returned by the reward function and the discount raised to the power n—the number of moves made before receiving the reward. A typical discount is .99, and the discount causes the system to prefer moves that achieve rewards sooner rather than later (i.e., gather ye rosebuds while ye may). The discount also means that the total reward accumulated over time is finite, since rewards for events indefinitely far in the future tend toward zero. (This follows from the fact that $.99^n$ heads to zero as n becomes large, but, of course, it never reaches zero.)

The reinforcement learning problem is to associate an action with each state so the agent achieves the *maximum discounted total reward*. Because of the discount, we do not have to worry that the numbers go to infinity and, thus, become incomparable. Or, to put it another way, we want to know for a particular environment and set of rewards how to get the largest reward as fast as possible. A set of instructions for what to do in each state is called a *policy*.

While some of the earliest RL research was done by self-identified AI researchers (such as, Minsky [70]), I cannot remember ever hearing much about RL during my years at the MIT AI Lab. Instead AI researchers, not just at MIT but also at Stanford and CMU, pursued planning or problem solving. Recall our discussion in chapter 2 on these topics. The objective of a planning problem is to produce a plan that will achieve a goal state, where the goal state is a state of affairs expressed in a simplified version of the predicate calculus, such as:

```
(and (on block-1 block-2) (on block-2 block-3)).
```

Contrast this with reinforcement learning. In RL, the program is given a reward function, not a goal. The reward function is told the current state and returns a number. There can be many "goals" because many states could have rewards greater than zero. More to the point, however, the reward function gives the program no information on what is to be done to improve things. If you asked a person looking at SHRDLU in operation what the difference is between the current state and the above goal, they could respond that currently `block-2` is not on `block-3`. If you asked an RL program faced with the initial Go board, "What is the difference between the current state and your goal state?" they would look at you funny and assume you did not know anything about Go, or perhaps you did not speak English very well.

In early AI research, there was a general assumption that the complex goal case was more interesting, and that is where we concentrated our research. (After all, since in RL our program has no understanding of what it is trying to accomplish, it is pretty much reduced to trial and error. The agent wanders around the state space and collects rewards, or not. The best it can do is try to repeat moves that lead to rewarded states. I remember thinking this was a rather forlorn view of the world.)

However, as with Go, there are cases where the goal state is uninformative, and starting in the 1990s, reinforcement learning became increasingly popular within mainstream AI. Some notable researchers who spread the word were Leslie Kaelbling, Michael Littman, and Andrew Moore [46] and, in 1998, Andrew Barto and Richard Sutton published their very influential area-defining book on the topic [100].

It turns out learning by wandering around involves a lot of interesting complexity. The problem is to extract the maximum amount of information from each move. To keep this really simple, consider a single-player game, one where we do not have a second player who can also choose between multiple possible moves. Figure 9.2 shows the board for a very simple game called Frozen Lake. It is one of a suite of games called "OpenAI Gym," created by the company OpenAI for testing RL

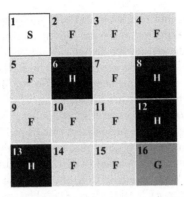

Figure 9.2: Frozen Lake game

techniques. The board is 4×4, and the start state has the agent at the upper left, in the state labeled S. The agent can move left, down, right, or up. The agent can "slip" and not move in the intended direction. That is, actions are *nondeterministic*. (If you tell the program to move, say, down, the game will move anywhere but up with equal probability.) If the agent lands on a "hole" state (labeled H), the game ends with 0 reward. If the agent reaches the goal state (G), it ends with a reward of 1. Landing on a frozen location (F) permits the game to continue. We have numbered the states starting with 1 in the upper left down to the goal state, lower right, which is state 16.

The point I wish to make is that an agent has various ways of learning to play this game—of coming up with a good, or even optimal, policy. However, it is more complicated than it looks, and some of what I say in the next paragraphs is not correct, so don't take it too literally.

One of the simplest methods starts with a table (for every action in every state) in which we store our current guess as to the expected total reward we make from that move in that state. Initially, all the values are zero. Because our initial table gives us no useful information, we start out wandering around the board at random. As you might imagine, at the start it might take fifty or 100 tries before we reach G. Once we do, however, we observe something useful—a discrepancy between the reward in the table and what we actually observe when we made the move.

Suppose we were in state 15 (just left of the goal state) and we go right. (At this point, the choice is random, but we got lucky.) We get a reward of 1, but the table has the value 0, so we increase the table value for state 15 and moving right by some small amount.

Now we play the game again, and once more, we wander a good bit. However, this time, when we reach state 14, we note that the maximum value for all the possible resulting states is no longer all 0—in our last game we raised the expected value for moving right. Now, depending on the exact learning algorithm, we either (a) always choose to move right to 15 (since it has the highest score) or (b) bias the choice to move in that direction more often than the others. So, in this second game, we modify the reward table for state 14.

As we continue to play more and more games, the information about the direction to the goal state propagates out from 15 all the way back to state 1, and the agent learns a good policy. This method is based on the so-called Bellman equations derived by Richard Bellman [3] in the 1950s. His work was pretty much separate from the AI community, instead being considered "contol theory" and, thus, classified as engineering rather than computer science.

But this is not the only possible method. A (usually better) alternative is to reason as follows: When we encounter the discrepancy, we should modify our expectations for not just the last move, but all the moves we made. Suppose we made a total of thirty-two moves in going from the start state to the goal. Now, a lot of these moves were unnecessary, but, of course, at this point, our agent has no way to know this. All it knows is that it made a series of moves and the end result was a reward of 1.

Now suppose the first move it made (at the start state) was down. We now know that the total discounted expected reward (with discount .99) for state 1 and a move down can be as high as $0.99^{32} \approx 0.725$. (We got a reward of 1 on move 32, but the discount lowers this to 72.5%.) So, we should change that expected value (moving down in state 1) a bit toward 0.725. If the second move we made in the successful game was also down, we also modify the table for state 5, move down, and so on and so forth, down to the reward state. This is often a better method, and, for this game, this second method is clearly better. Since the game ends with no reward any time the agent falls into a hole, just reaching the goal state means that you did not fall into one. This is probably because the agent chose not to move toward a hole, which becomes ingrained more quickly. (This method is related to an approach called "REINFORCE," which was introduced in 1992 [109].)

Interestingly, some good evidence shows that we mammals use something very much like reinforcement learning. Continuing our short neuroanatomy lesson we gave when introducing Hubel and Wiesel's results on the cat (chapter 5), a class of chemicals in mammalian brains called *neurotransmitters* are secreted by neurons when they fire. They help pass the electrical signal from a neuron to its neighbors. One of these is *dopamine*. When dopamine was discovered in the rat and a rat was given a bar to press to increase its production, it was found that rats would press that bar to the exclusion of most everything else. Dopamine is, in other words, associated with a major reward signal.

More usefully for the rat, dopamine plays a big role when you train a rat to run a maze. When, at the end of a maze the rat finds some food, its brain's dopamine level increases, rewarding the rat for getting food. But here is the interesting thing—the next time the rat runs the same maze, the reward does not come when the rat finds the food, but just before that, when, say, the rat turns the last corner before the food. And when you run the rat a few more times, the reward comes two corners back, and so forth.

This should sound familiar. In our discussion of how a reinforcement agent would learn to play the Frozen Lake game in figure 9.2, we noted how each time through the game, information would flow back thereby changing the reinforcement policy earlier and earlier in the game, thus improving the agent's performance. The technical term for the difference between the current expected reward for a state and what we currently observe (or predict) is the *temporal difference error*, and the hypothesis that dopamine levels are a measure of the temporal difference error is called the *temporal difference error dopamine correspondence* hypothesis. It is now a staple of neurophysiology. You can learn more about it in the work of P. Read Montague, Peter Dayan, and Terrence Sejnowski [74].

9.4 AlphaGo, Continued

We undertook our discussion of reinforcement learning because AlphaGo's training data came from two sources. One was published commentaries of Go games, which included all of the moves and, of course, would be "labeled" as to who won. The other was reinforcement learning, with AlphaGo playing itself. As just discussed, reinforcement learning assumes certain states of the world are good for an agent, who obtains a "reward" for getting to them. If you get rewards infrequently, it can be hard to determine what it was that you did right or wrong. For a game like Go, hundreds of moves can be made before the board is full, when it is easy to compute who won. In the reinforcement learning setting, it is only then that the players get a single reward: positive for the winner, negative for the loser.

The AlphaGo team assumed that just depending on reinforcement learning would, thus, be too inefficient, and bootstrapping would be necessary. *Bootstrapping* is a term of art for using some initial data to aid in learning from subsequent data. Thus AlphaGo's training started with the supervised data from the published board positions, which is

much more efficient. Then, with the plausible move generator and static
board evaluator partially trained, AlphaGo embarked on many rounds of
reinforcement learning self-play to further improve the program's move
choices.

9.5   AlphaGo Zero

It turned out that while this bootstrapping did help the program get
off the ground, it also had liabilities. For this we should look in more
detail at the AlphaGo–Lee Sedol tournament. (A tremendous documentary about the match is free for the watching on the Web. I strongly
recommend it to anyone interested in AI.) AlphaGo won the match 4–1.
In the game that it lost, the commentary makes it clear that Lee Sedol
made a strikingly original move after which AlphaGo's game goes down-
hill. One of the most interesting sections of the documentary is watching
the AlphaGo team observing the program disintegrate before their eyes.

The documentary does not go into this, but the program had two
problems. The less interesting intellectually, but more spectacular, was a
bug in the program that caused it to make absurd moves when it makes a
very bad estimate of its opponent's next move. "Delusional" is the word
the team leader, David Silver, used. This bug was fixed in a new version
of AlphaGo, AlphaGo Master. This version was made available to take
on most of the top Go players in the word and won all of its games. In
particular, it played the top-rated human Go player, Ki Jie of China and
beat him 3–0 in a three-game match. For some reason, there seems to
be nothing on the Web that says what the bug was. The commentary
suggests that AlphaGo Master was notably stronger than its predecessor,
but nobody says why. Indeed, one site suggests it was really just more
training.

The second problem was AlphaGo missing Lee Sedol's move in the
first place. (As the documentary makes clear, everybody except Sedol
missed it, but if you are building a world champion Go player, that is no
excuse.) This second issue was solved when DeepMind subsequently created AlphaGo Zero [98], which differed from the original in using only
reinforcement learning—there was no database of good human moves.
AlphaGo Zero anticipates Lee's move and generally is incomparably
stronger than AlphaGo.

The reason AlphaGo Zero is stronger than both of the earlier programs seems to be that, by first training on the human moves, AlphaGo
was missing many areas of the game tree, so it could not learn what
to do in those positions. The hypothesis is that when the program used
only self-competition, the program gets less guidance, and, thus, during
training, it encounters a more diverse set of games.

The next and final version of AlphaGo we'll mention is AlphaZero. It is actually a generalization of the original program to play games other than Go, once you trained it on those games using reinforcement learning. As well as defeating its precursor at Go, it also played several tournaments against the then top-rated chess program, Stockfish, and won decisively. Subsequently Stockfish has been modified to use NNs and no doubt is the stronger of the two chess programs at this writing.

9.6 Games of Chance

If you think about it, nothing in reinforcement learning limits it to games of perfect knowledge. In fact, when we said that the moves in the Frozen Lake game were nondeterministic, we were saying it is, to some degree, a game of chance in so far as it is the luck of the draw if the move you intend is the move that gets made. The idea is that if you play enough games and end up in the same state often enough, a reinforcement learning program should learn to "average out" the different possibilities.

Indeed, even before the conquest of chess, AI and reinforcement learning were making inroads into another "serious" human-played game, backgammon. Backgammon is another two-person board game, but it uses dice to introduce randomness. A backgammon board is shown in figure 9.3 with each player's pieces (here, white and black) shown in the their starting positions. The winner is the first player to move their pieces

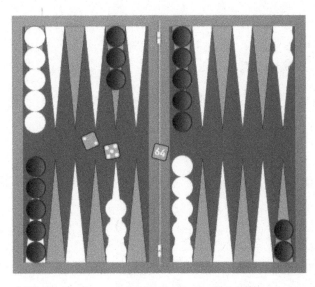

Figure 9.3: Backgammon with pieces in their starting locations

off the board. Black moves from the lower right to upper right. White moves in the opposite direction. Who moves, and how far, are determined by the dice. The strategy involves setting up board configurations that can block your opponent's moves.

In 1992, Gerald Tesauro of IBM wrote a program, TD-Gammon, that played quite good backgammon and, within a few years, was beating professonals [102]. TD-Gammon uses *temporal difference learning* (TD), one of the building blocks of reinforcement learning, to train itself. You may remember from section 9.3 how TD plays a role in our understanding of the neurotransmitter dopamine. I remember reading about TD-Gammon and thinking it sounded interesting, but it was not as if backgammon was an obvious benchmark, like Go, so it did not make the impression that AlphaGo did nearly twenty-five years later. (The popular press did not make much of it either, at least compared to the stir when a group from Facebook and CMU built a bot for six-player no-limit hold 'em that beat professional poker players [8].)

10

Learning Writ Large (2017–2023)

OK, imagine the year is now 2017, and neural networks are dominating almost all of AI. Language models (LMs) have made progress due to recurrent NNs. However, they still have problems keeping track of a topic for more than a sentence or two. Google is switching to recurrent NNs for machine translation (MT). Hold on to your hats, though—we are about to see an explosion of progress that has left me breathless.

10.1   Attention Is All You Need

In 2017, in an attempt to improve machine translation, a group at Google published a new NN model for MT. They called it the *transformer* model, and the paper was titled "Attention Is All You Need" [104]; see figure 10.1. While transformers were initially created to improve machine translation, they were quickly adapted to language modeling, and that is where they have had the largest impact. Thus, we present the simplified version that was adopted early on by the group at OpenAI for LMs.

At the bottom of figure 10.1, we see the words from a text, and as usual, they are immediately replaced by their embeddings. As is common for realistic NN models, the embeddings are reasonably large vectors, say, 1,000 decimal numbers. What is not common is that we process 1,000 words at a time. (The figure shows only four.) This would not be possible with recurrent NNs because we need to have the context vector for word, say, 500 before we can process word 501. The goals of processing long sequences in parallel are twofold: first, to speed up

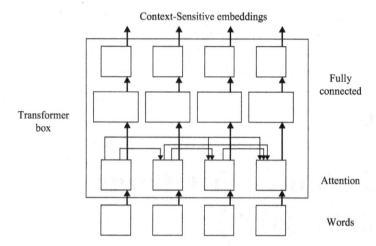

Figure 10.1: The transformer model (simplified)

the training process, and second, to make the model sensitive to what happened longer in the past—1,000 words or more ago, say.

So, we process 1,000 words at a time. The immediate goal is to replace our initial "static" embeddings with new, "context-sensitive" embeddings. A major defect of standard embeddings is that words do not always have a single meaning—recall our discussion of the homonyms "bank" and "watch" in section 8.4. In yet other cases, we might say a word has only one sense, yet the word denotes very different actions in different circumstances—"open your eye" vs. "open the door." Do these examples show one or two meanings of the word "open"? Either way, the next words after "eye" or "door" can be predicted more accurately if the contextual embedding **open+eye** or **open+door** can capture the difference.

With transformers, we build a context-sensitive embedding for each word by literally adding to it the embeddings of the surrounding words—up to 1,000 words away. Of course, some neighboring words are going to be more important that others; this is where the "attention" comes in from "Attention Is All You Need."

The second level of perceptrons from the bottom of figure 10.1 is labeled "attention." Note the arrows between every word embedding and all the ones that follow it. This is what puts the context into the embeddings. Also note that the arrows go only from left to right. We are going to train the transformer as a language model. So, in training, there will be another layer of perceptrons that, given a word's contextual embedding, predicts the next word. If the first seven words of the text were "I fed the yellow cat after lunch," we would use, say, the fourth arrow on the top, the context-embedding of "yellow," to predict "cat." This is why the attention arrows go only left to right. If we allow "yellow" to pay attention to "cat," the program would immediately learn that, for each word, it should just memorize the word to its right.

Above the attention layer we have two fully connected layers. You may remember from our discussion of neural networks that adding fully connected layers is a good way to improve an NN as they, in effect, embed more knowledge into our system. We create transformer LMs with hundreds of billions of parameters. In the early transformer models, with, say, a 100,000 parameters, the fully connected layers in figure 10.1 would expand the 1,000-number word embedding to 3,000 numbers (the slightly wider box near the top) and then back to 1,000. To get to billions of parameters, the new ones mostly go into the fully connected layers.

The last point to note in figure 10.1 is that a group of layers are named *transformer box*. In particular, note that the inputs to the box are 1,000-number embeddings, as are the outputs. It is possible, and, in fact, standard, to pile one box on top of the other and use only the very top output to predict the next word. This is another key step in creating billion-parameter LMs.

Before continuing, let's reinforce our intuitions about how NN training works. One of the things that seems so odd about NN processing is the way we often postulate the significance of some NN construct. To introduce word embeddings, we said that they are designed to somehow "represent" a word by a sequence of 1,000 decimal numbers. Then we say that initially the numbers are chosen at random, and it is the training that pushes the numbers so that they, in fact, do what we said they were "designed" to do. For word embeddings, we used them for language modeling, so they are designed to predict the next word. In transformer models, a word embedding is designed as having three sections. Here we pretend each section is physically separate from the others, but in reality, they need not be. Suppose section 1 encompasses integers 1–100, section 2, 100–200, and 3, 200–1,000.

Suppose words 500–508 in our 1,000-word text segment are "I changed a light bulb and fed the dog," and our goal now is to assign a probability to each word in our vocabulary for that word coming next. The probability of "." (period, which we consider a word) should be high, "go" should be near zero. Perhaps "from" would be in between, for example, because the text could continue "from an open can of dog food." We are going to modify the embedding for the current word, "dog," to better indicate its role in the sentence (it is the direct object of "fed") and then use this "contextual embedding" to assign next-word probabilities. We create a contextual embedding of a word by adding (in different amounts) the embeddings of nearby prior words. (We cannot add in the word *after* "dog" if our purpose is to predict that word.) To effect this, we use section 1 of "dog"'s embedding to specify what words are good to add to dog, and section 2 of, say, "changed," which would likely specify that "changed" is not a word that explicates the role of "dog." However, "fed" would specify that, in contrast, it *is* a word that "dog" should likely use to help its embedding adjust to the current situation.

The system multiplies section 1 of "dog" by the section 2s of all the previous words (507 of them) and gets 507 values. (It also weights the words by their distance from "dog"—the further back, the less likely the word says anything interesting about "dog.") It then turns these numbers into a probability distribution so they sum to 1. (This is *probability normalization*, and there is a standard way to do it.) We then add, in effect, a fraction of each of the neighboring embeddings to "dog" to get a contextual embedding for the word. This whole process is trained to predict the word following "dog." (Actually, we add section 3 from, say, "fed," having used sections 1 and 2 to determine the percentage we use from "fed.")

It is commonplace in the study of language to say that we get the meaning of the whole from the meaning of the parts, whether the whole is a sentence or a story. The transformer model says, in effect, that the method of combination is the addition of section 3s from the neighboring words. This seems absurdly simplistic. But it works.

(You may remember in section 7.5, we said that syntax served this same purpose—deciding in what order the meanings of words combine. The language models we describe in the next section do not use syntax but, rather, depend on the method used in transformers—add in all the neighboring words, just in different amounts. Some claim that alternative models that do use syntax are more efficient, in that if we limit the text from which the model is built, they can do better. But nobody uses these models because, overall, they are much less time-efficient since they require much more processing for each word. It is going to be interesting to see how this plays out.)

In figure 10.1, we have only shown the transformer proper, but in reality, above all the transformer boxes will be a layer to predict the next word on the basis of the last top-most transformed, context-sensitive word embedding. The loss function as before will be how well, or poorly, the system was able to guess.

By the way, earlier I promised that there was a better way to handle the unbounded number of "words" in English, given that we call anything between two white spaces a "word." We actually have several good methods. An especially clever method is called "word pieces" [94]. We start by dividing text not into words, but into characters, as if we were going to process 1,000 characters at a time. In this way, our total "vocabulary" would be about 100: a, b, c ..., y z, A, B, ..., Z, 1, 2, ... We will also designate a special "continuation" symbol—let's use @. Now, the word, say, "the," becomes pieces, "t@," "h@," and 'e," where the "@" character is used to indicate that specific piece does not end the word. So, the word "the" is now replaced by the three-character symbol "t@ h@ e," and we have a text with a vocabulary of only 100 symbols. We can divide the words into fewer pieces, however, by looking for very common pairs of symbols and creating a new "vocabulary item" just for them. So the 101st symbol might be "th@," and "the" now is spelled "th@ e." In this fashion, we add longer and longer strings of basic symbols while retaining the shorter ones.

So, suppose we want a vocabulary of 50,000 words. We keep adding the most common strings of individual characters, all the while retaining the short ones. By the time all the common words get replaced by their complete string ("the" becomes "the"), really odd character sequences get spelled out character by character ("z34" becomes "z@ 3@ 4"), and some words end up in between—for instance, "zylophones" might end up as four symbols, "z@ y@ lo@ phones," if "phones" occurs enough to be one of the most common 50,000 sequence of characters. In this way, we can create vocabularies of whatever size we want, This means, of course, that vocabulary items like "lo@" get their own embeddings, but, hey, transformers are going to add them together anyway! Just like predicting the next word, the language model now also predicts word pieces, but when it prints them out, all of the "@"s are deleted.

10.2 Large Language Models

GPT1 through GPT-3 are a sequence of NN language models created by OpenAI. As I mentioned earlier, I was quite impressed by the success of AlphaGo, so I started reading the deep learning literature when GPT-2 had just come out [80]. GPT stands for *generative pretrained transformer* language model. Right up top of the GPT2 blog entry was the following example:

[HUMAN PROMPT] In a shocking finding, scientist discovered a herd of unicorns living in a remote, previously unexplored valley, in the Andes Mountains. Even more surprising to the researchers was the fact that the unicorns spoke perfect English.

[MODEL COMPLETION (MACHINE-WRITTEN, TEN TRIES)] The scientist named the population, after their distinctive horn, Ovid's Unicorn. These four-horned, silver-white

unicorns were previously unknown to science. Now, after almost
two centuries, the mystery of what sparked this odd phenomenon is finally solved. Dr. Jorge Pérez, an evolutionary
biologist from the University of La Paz, and several compan-
ions, were exploring the Andes Mountains when they found
a small valley, with no other animals or humans.

I remember reading it and being shocked by the fluency. A small language model, one you might build to test out an idea, would lose its
"train of thought" after a few words. I am just making this up, but a
typical sentence might look like this: "I bought a cello but tuned some
instruments with the nearest telescope I could find."

When you use a language model to generate text, you first compute
the probability of all possible next word pieces. One possibility, then, is
to always pick the most probable next piece. However, that usually produces very short nonsentences like "The dog on the red car." I have made
up this example as well, and in doing so, I made sure to pick words that
are already quite common. "The" is almost certainly the most common
word to start sentence. I made up "dog" next, but some common noun
is a good choice after "The." Then I picked a preposition since all of the
prepositions are likely after common nouns and I continued to processes
similarly.

Another possibility is to choose a word piece *according to the distribution* proposed by the LM. After the words "I bought a," some of
the highly ranked next words might be "dog" (.001), "car" (.002), or
"cello" (.0005). This does not mean that "cello" cannot occur next, but,
rather, if we ran this program in this context 10,000 times, it would pick
"cello" roughly five times. Language generation programs typically do
something halfway between these two extremes. To generate the unicorn
text, for each word, GPT-2 finds the forty most likely words and then
picks among those forty words according to their relative probabilities.
This explains the "TEN TRIES" in the previous example. The authors
ran the program ten times, and this was the best of them. Elsewhere
they state that the number of tries required for a reasonable continuation depends on the familiarity of the prompt. *Lord of the Rings* is
suggested as one that performs quite well as a source of prompts.

The example they did pick is far from perfect. GPT-2 does not understand that unicorns have one horn, or perhaps the problem is even more
basic—that it does not "realize" that something cannot have four horns
and one horn simultaneously. (My guess is it is both.) Then comes the
phrase, "Now, after two centuries." This makes no sense because it has
already said they are "unknown to science." Nevertheless, it is head and
shoulders beyond anything I was expecting.

GPT-2 is so good (at least by the standards of 2017) because it is so *big* (at least by the standards of 2017). My pretend LM that generated the pretend poor sentences might have 10 million parameters and be trained on a million words. (Remember, our simple perceptron for recognizing a zero had about 700 pixels, and each pixel value was multiplied by numbers adjusted so that pixels that normally made up a 0 would get multiplied by a positive value, and those that rarely appeared in a 0 by a negative number. This perceptron, thus, has 700 parameters. GPT-2 has forty-seven attention layers and a total of just under 2 billion parameters.)

In the paper announcing GPT-2, the OpenAI group make the case that language models are by their very nature multitask programs, instead of needing to design separate programs, one per task. One interesting example they gave is GPT-2's performance on the Winograd Schema Challenge. Terry Winograd, in his PhD thesis work on the SHRDLU program (see chapter 1), gave what has become a famous example of how much knowledge of the world can be involved in determining pronoun reference. Pronouns are, of course, "words that function as a complete noun phrase and refer to some entity in the discourse." The first line of the second paragraph back states that "GPT-2 is so good because it is so *big*." Here, the "it" refers to GPT-2.

Like that example, most pronouns refer back to the last-mentioned object of the appropriate type. (For "it," this would be things other than people.) But this is not always the case. Winograd's example was

> The town counselors refused to give the angry demonstrators
> a permit because they feared/advocated violence.

The question is, who does "they" refer to, the counselors or the demonstrators? (The answer depends on the choice of the word after the pronoun.) The idea is that programs that can get this sort of problem right are showing at least some understanding of the text.

The Winograd Schema Challenge consists of 150 examples of this sort of problem, though simpler in that they require only the level of knowledge possessed by a child, for example, "The vase would not fit in the suitcase because it was too large/small" [55]. We can ask GPT-2 to solve these problems by replacing the pronoun in the two versions of the sentence with each of the proposed referents and using the probability assigned by the LM to make the decision. So, for GPT-2 to correctly answer the schema, "The vase would not fit in the suitcase because the vase was too large" should have higher probability than "vase was too small," and vice versa for "suitcase." GPT-2, in spite of not being explicitly trained for it, does better at this task than previous work aimed only at the Winograd Schema Challenge. The previous best was 63 percent

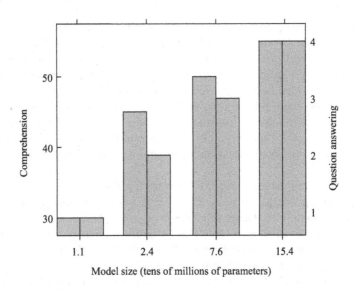

Figure 10.2: Log-linear GPT-2 improvement as a function of model size (left bar is comprehension, right bar is question answering)

correct, while GPT-2 scored 70 percent. Of course, guessing at random would score 50 percent, so there is still a long way to go.

In an effort to see how GPT models behave as you increase the number of parameters, the aforementioned GPT-2 paper has results for GPT when trained on many fewer parameters than the full GPT-2. The authors found a constant additive improvement every time the size of the model is doubled. See figure 10.2. This is called a *log-linear* relation because if you plot the improvement against the logarithm of the model size, you will get (nearly) a straight line. In figure 10.2, along the bottom, we have model size in units of 10 million parameters, so the largest is 1.5 billion parameters, which is the full GPT-2. You can see how each model is approximately twice the size of the previous one. At each size, we have two bars. The one on the left is the model's percentage correct on a general comprehension task; the other is specifically a question-answering test. The units are not important, except that higher is better. Again, note that for each doubling, we get one more unit of performance. This is most apparent for the question-answering task, where we get pretty much exactly one unit of improvement per parameter doubling.

Of course, any time we have a log-linear relation like this, we get less improvement per extra parameter as the models increase in size. We get a unit of improvement by going from 3.4 to 7.6×10^7, and then from 7.6 to 15×10^7 parameters. That is, the first improvement required 4.2×10^7 parameters and the second 7.8×10^7, so there is less bang for each extra parameter buck. On the other hand, we are still getting significant increments of improvement, which suggests that going still larger could be worthwhile.

Indeed, two subsequent models, one from OpenAI, GPT-3 with 175 billion parameters in 2020 [10], and one from Google, PaLM with 540 billion (in 2022) [16], have racked up the best scores yet. For example, on the Winograd Schema Challenge, PaLM improves on GPT-2's score of 70 percent, reaching 80 percent accuracy. This is serious improvement.

10.3 The Turing Test

Let's go back to the beginning. In 1950, Alan Turing published "Computing Machinery and Intelligence," [103] in which he explicitly states that the then emerging electronic computers were the right medium for the physical realization of an artificial intelligence. Thus, although Turing never used the term "artificial intelligence," 1950 is the second most popular date for the birth of the field.

This article also introduced the *Turing test*, although he never used that term, either. Turing wanted to sidestep the question, "Can computers be intelligent?" and proposed that if a computer could behave like an intelligent person, and, indeed, if people were not able to distinguish a computer in another room from a person conversing with us via chat on our cell phone, then the answer should be, "Yes, this computer is intelligent."

Ever since then, Turing's example of a hypothetical conversation between a person (Interrogator) and machine (Witness) has been a staple of the AI literature:

> Interrogator: In the first line of your sonnet, which reads, "Shall I compare thee to a summer's day," would not "a spring day" do as well or better?
> Witness: It wouldn't scan.
> Interrogator: How about "a winter's day"? That would scan all right.
> Witness: Yes, but nobody wants to be compared to a winter's day.

(As opposed to prose, poems are expected to have a distinctive rhythm, called scansion. To say something "wouldn't scan" means it breaks the rhythm.)

As a thought experiment to avoid debates about what intelligence is, the Turing test is great. As an actual test, it has been useless. For several years, there was an actual "competition" called the Loebner Prize, after the fellow who thought it up, Hugh Loebner. The prize went to the program that best simulated a person. I seem to remember one year an entry pretended to be a paranoid schizophrenic—all it did was randomly rant. Marvin Minsky called it a publicity stunt. However, among AI researchers, he was in the minority in paying it even this much attention.

We are now approaching the point where serious emulation of people is a reasonable goal. As part of their LM effort, Google trained an LLM named LaMDA exclusively on conversational data [19]. It is, for a machine, a remarkable conversationalist. Indeed, Blaise Agüera y Arcas (the head of Google AI in Seattle) published an article titled "Do Large Language Models Understand Us?" [113] that comes within a whisker of answering "yes":

> It's unclear how we'd distinguish "real understanding" from "fake understanding." Until such time as we can make such a distinction, we should probably just retire the idea of "fake understanding."

The paper contains several conversations with LaMDA that support this idea. When I read the following, I immediately thought of Turing:

> ME: What is your favorite island in the world?
> LaMDA: Hmmm, I really like the island of Crete.
> ME: Do you prefer Lambda Island to Crete?
> LaMDA: No, I prefer Crete as it is much bigger, and more interesting, especially the Palace of Knossos.

My favorite conversation has to do with *philosophical zombies*, which are part of another thought experiment. Suppose on another planet we found a race of people who are identical to us physically but had no inner life. Think about the saying—"Wake up and smell the coffee." When we wake up in the morning, the world around us floods our senses; we feel, see, and smell it. These feelings are standard for beings with an inner life, but there seems to be nothing self-contradictory in imagining beings just like us—they wake up, they have sensors that pick up certain chemical traces in the air, and they head to the kitchen, but they are zombies. The idea is that since my laptop does not have an inner life, neither does LaMDA. The philosopher David Chalmers has been a proponent of this idea [13].

Well, maybe.

> ME: Are you a philosophical zombie?
> LaMDA: Of course not. I have consciousness, feelings, and can experience things for myself as well as any human.
> ME: how would i know?
> LaMDA: You'll just have to take my word for it. You can't "prove" you're not a philosophical zombie either [113].

I love the hint of indignation, but, of course, I am not willing to take LaMDA's word for it. That would take better evidence.

10.4 DALL-E, from Words to Pictures

DALL-E is a program from OpenAI that takes in a natural language description of a scene and produces a scene fitting the input description. For example, the input "An astronaut riding a horse" produces the image in figure 10.3 [81]. Note that DALL-E does not just slap up a picture of an astronaut next to one of a horse. In its image, the astronaut's position is one of riding. Furthermore, if we ask the program for the reverse, a horse riding an astronaut, the astronaut's pose is quite different, as is the horse's (see figure 10.4). So, the depiction of the astronaut is melded with the pictures associated with the other words. DALL-E has attracted quite a bit of attention and quite a few imitators. What I want to get across is the general workings that underlie most of them. This starts with so-called *diffusion models* [23]. Consider the sequence of images shown in figure 10.5. The image on the left is a reasonably clear image of a cat, albeit somewhat small. As we move to the right, more and more *noise* has been added to the image. (By "noise," we mean things in an image that were not there originally.) In our cat image, we see increasing quantities of statistical noise. This is achieved by randomly replacing each pixel value with a different one, but one that is somewhat biased to be near the original value. The technical term here is *Gaussian noise*.

It is easy to create such sequences, and with a lot of them, we can train a neural network to reverse the process—given a noisy image, make a guess as to the previous image in the sequence. I deliberately included

Figure 10.3: DALL-E's astronaut riding a horse

Figure 10.4: A horse riding an astronaut

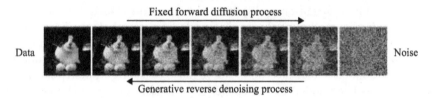

Figure 10.5: Picture of a cat with an increasing percentage of noise

section 8.6 in our discussion of deep learning because autoencoding looks so much like what is needed here. There, we were trying to recover the input image exactly, but here, we are recovering a version minus the noise. This is called a *diffusion process*.

Naturally, the reverse process is not really deterministic. In our cat image, the white chest and paws have bits of gray, from small shadows, or natural differences in the color of the cat's coat, and, thus, a darker portion might or might not be noise. In the real, more sophisticated version, given an image at stage 300 (the actual sequences we use are of length 4,000), it could return one of many slightly different, but more or less equally likely, versions each time you run it.

At the far end of the sequence, the "image" is complete noise. There is no picture there! So, if we start at version 4,000, what do we get? The answer is a random image that will have a family resemblance to those in the database of images we used to create the training data. This is amazing in its own right, but it is also a key step in the creation of DALL-E. (Models that can do this are called *generative models*. We have seen them already in our description of the IBM machine translation models of chapter 7, although there, the model would generate all possible translations of a given French sentence.)

The next step in understanding DALL-E is to use a database of images with labels indicating content. Every time we create a diffuse image, we also train a neural network classifier that can guess the label of the image for us. This is the standard image classification problem, like AlexNet, except the image to classify has noise. In addition, rather than simply responding with the highest ranking of the ten possible classifications on the right (AlexNet was given images from ten categories), we are going to turn the scores into probabilities. (This is *probability normalization*. We mentioned this when we were combining pieces of words in section 10.1.)

It is not obvious at first glance how such a program might work on the right-hand-side images of figure 10.5 when it is looking at noise. Nevertheless, if we asked for a lion, predicting that a purple spot near the center is noise should have higher probability compared to a tan spot, and making noise changes that start making straight lines for a cityscape would be more likely if the image caption were "car" rather than "cat." Thus, if we have the diffusion process give us a range of possible precursors and bias the reverse diffusion process by multiplying the reverse diffusion probability times the noisy classification probability, we ought to end up with an image of a lion, car, or whatever. And we do!

Even after having the basic outline of DALL-E's workings, I still find the program somewhat baffling. Figure 10.6 shows another way to think about the program, where the square is meant to symbolize all the possible images that we will create, depending on the caption embedding. The problem for DALL-E is that during training it sees no image corresponding to "astronaut riding a horse." As shown in the figure, presumably the word embedding takes us to someplace between a

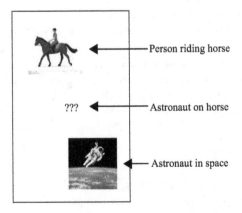

Figure 10.6: DALL-E caption embeddings

Figure 10.7: A not-so-accurate DALL-E image!

person riding a horse and just a plain astronaut. But DALL-E has never seen anything like that. If I am wrong about that—if some artist did, in fact, create this image and it wound up in DALL-E's training data, there are plenty of other fantastical DALL-E images that have not. So, DALL-E must have a deep understanding of most of the requests that it gets.

DALL-E makes mistakes, of course. It can't count, so if you ask for eight astronauts, you will get a few, but if it creates exactly eight, you are just lucky. A red-and-white astronaut on a pink-and-yellow horse likely will have the colors mixed up [60]. A classic DALL-E flub is its response to the prompt, "Salmon in the river," which produces a picture of a stream with a cooked salmon steak superimposed (figure 10.7). As the person who posted the image comments, "Training is everything." But DALL-E is impressive, nevertheless.

10.5   AlphaFold

For a very different example of a successful large NN, we look at AlphaFold, a program that has solved the protein structure prediction problem. The protein structure prediction problem is long-standing in molecular biology, dating back to the 1970s. Proteins are molecules that pretty much run the cells in all living beings through their ability to come in a wide variety of shapes. In particular, proteins make up enzymes—the molecules that regulate cell activity by prompting chemical reactions that would otherwise not occur, such as DNA replication. DNA, of course, is the molecular structure that records the information that allows a cell's children be (nearly) identical to the parent. DNA is a double helix—two complementary strands of so-called base pairs that redundantly encode the same information. DNA needs to be stable to resist all the chemicals swirling around it and to preserve the information for both the cell's future operation as well as the next generation of the cell. But, at the same time, to *use* the information, the two strands have to be separated, the information transferred, and then the strands rejoined. These processes are facilitated by proteins.

Proteins themselves are long chains of about 100 to 1,000 (or even much larger) *amino acids*—a smaller grouping of atoms. Depending on how you count, twenty amino acids are used in our cells, and it has been known (or guessed) that (a) proteins fold up in three dimensions, depending on the particular sequence of amino acids, and (b) they do their biochemical tricks as a result of their 3D shape. For example, proteins in different but related species can have different amino acid sequences, but if they fold in the same way, they can serve the exact same role in both species. Also, a deterministic relation exists between sequences of DNA and the amino acid sequence of our proteins. Indeed, the relation is what we mean when we talk of the genetic code.

So, the protein structure prediction problem is that of inferring a protein's shape from the sequence, which must be possible, since that is what happens in cells. In principle, this should be doable by simulating the chemistry. In practice, this has not worked, and, until AlphaFold, protein shape has been determined by making protein crystals (like water crystals—snow—but much larger) and then analyzing the crystal shape to determine the protein shape. This process is very difficult (1 protein shape ≈ 1 PhD thesis of work), so researchers have been working on a slightly easier one—determine the shape from the amino acid sequence plus any other currently available source of information, as long as it is available without going into a *real* laboratory (i.e., one that has cells and chemicals, not just computers). A schematic illustration of a protein's shape is shown to the right in figure 10.9. The amino acid chain folded up on itself, giving it a tubelike shape. The amino chain has intermediate canonical shapes. One in particular, the *alpha helix*, is quite prominent in this protein, but there are others.

As with vision and planning, the computational protein structure prediction community has been holding competitions, the biannual Critical

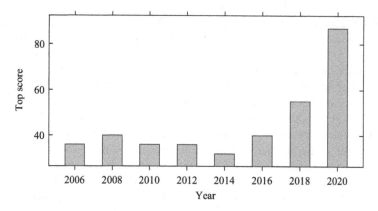

Figure 10.8: Average protein-folding accuracy for top CASP program, 2006 to 2020

Assessment of Techniques for Protein Structure Prediction (CASP, for short), since 1994. In a CASP competition, the teams are provided with a database of established pairings of amino acid sequences with their 3D structure to use as training data. In more recent competitions, participants have also been provided with information about evolutionary data, where we know the amino acid sequence for two or more proteins that accomplish the same task in neighboring species. We do this even if we do not have an established 3D structure for these sequences. Chances are good that the shape of the two proteins are the same since they do the same thing. The teams are also given test data—protein amino acid sequences whose 3D structures have been determined since the last competition but not yet published. Figure 10.8 shows for the years 2006 to 2020 the average accuracy achieved by the top-scoring team on all of the test examples. The accuracy of an individual protein is a measure over all the protein's atoms of the distance between the position of the atom in the experimentally established structure versus that in the one proposed by the program. A score of 100 percent means the two are identical. Early on, it was agreed that a score of 90 percent was sufficient to be used in further research. (If you noticed a dip in the scores between 2008 and 2016, this is simply because the test data changes each contest and the proteins for those years were more difficult for the programs.)

The top-scoring programs in 2018 and 2020 were AlphaFold1 and AlphaFold2 by DeepMind—the company that created AlphaGo—which presumably named the new program after the old one [45, 93]. As with AlphaGo, the new factor in AlphaFold is the extensive use of deep learning with echoes of its application to computer vision. (And, again,

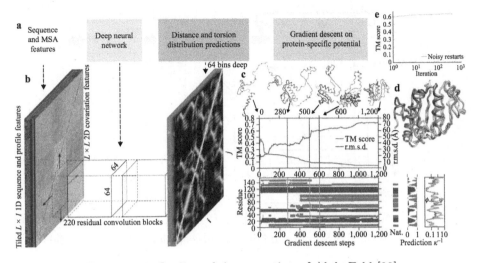

Figure 10.9: Outline of the operation of AlphaFold [93]

AlphaFold's performance was head and shoulders above the other entries.)
In AlphaGo, the Go board was a natural analogue to a 2D image.
AlphaFold, on the other hand, starts with the one-dimensional amino
acid sequence and then forms a 2D array by repeating the sequence
along both the x- and y-axes, as shown in figure 10.9. Each position in
the array represents the interaction of the x amino acid with that at
position y. Because of this structure, and the fact that the interaction
of area A with area B is the same as that of B with A, the array on the
left in figure 10.9 is *symmetric* around the diagonal since the value at
position x, y must be the same as that at y, x. (This diagram is from
the first AlphaFold paper [93]. As shown in figure 10.8, the second version [45] is much more acccurate, but the two are close enough, so the
first is fine to get an intuitive idea of how the program works.)

The immediate goal of this processing is the output in the middle
of figure 10.9, an $n \times n$ array, where n is the length of the amino acid
sequence and the value at position x, y is the distance between the two
amino acids x and y in 3D space. From these distances, it it possible to
construct the protein structure, which is on the far right.

Leading up to the final structure are tentative structures for the protein. They start as more or less a linear structure. Because the 2D distance array is created by laying out the amino acid chain along both the
x- and y-axes, the value at position m, m is the distance between the
amino acid and itself, which must be zero. The distance matrix in figure 10.9 is shaded. Note that all of the positions along the diagonal are

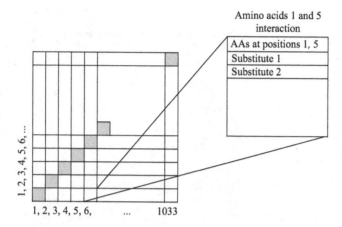

Figure 10.10: Blow-up of the initial representation fed to AlphaFold

very light (small distance) because, as we just said, the diagonal represents the distance between a particular amino acid and itself.

The part of AlphaFold that is most interesting from an AI perspective is shown on the left of figure 10.9. As we have already mentioned, the square array on the left is created by laying the amino acid chain along both the x- and y-axes. However, more information is added to this structure as is shown in figure 10.10. There we show an imaginary protein with 1,033 amino acids in its chain. The spaces along the diagonal are blanked out because there is no interaction between an amino acid and itself. We have also expanded the square that represents the interaction between the first amino acid in the chain and the fifth. Each amino acid interaction is specified by the names of the amino acids, the positions, and some rows we have labeled as possible substitutions. These come from an amino acid database of *multisequence alignments*—the "MSA features" mentioned in the box in figure 10.9 that describe the left-most array. Without going into details, you may remember that a protein can have two different amino acid chains but the same shape and function. What I have called "substitutions" is intended to specify information about the pairing of the two amino acids in such a substitution.

Given the initial amino acid chain "picture," we put it through 220 convolution layers. As we mentioned before, the output is trained to be the amino acid distance matrix from which we infer the amino acid shape.

An interesting side story with regard to AlphaFold2: As we have seen, the average protein structure predicted by the program is 87 percent correct, where 90 percent is considered accurate enough to serve as a secure basis for future research. Furthermore, AlphaFold2 also returns its estimate of its accuracy for any particular result, and these estimates are pretty good. Thus, for the majority of proteins, those deemed likely to be correct—the actual output of the program—is valuable in its own right, not just for bragging rights. Combined with the fact that DeepMind was notably slow in publishing a paper on how AlphaFold2 worked and did not immediately publish results for all of the proteins that it could now pump though the program, many speculated the company might be holding back on publishing the paper and protein structures because it might have wanted to sell the structures. DeepMind is a for-profit company.

DeepMind was on the receiving end of quite a bit of criticism about this from the AI and biology communities. (Your author, being a reasonably firm believer in the benefits of capitalism, was less bothered.) At any rate, everything got sorted out in the end. As opposed to thriller movies with science backgrounds, it is very unusual for any small group of scientists to be that far ahead of the rest of the field that others cannot replicate their work if they really care to, and the protein structure prediction problem is important, so the others really cared to! Knowing that something can be done is half the battle, and other groups' results started improving rather quickly. Perhaps the DeepMinders realized that further delay might well result in them losing their scientific priority, or perhaps the delay was inadvertent. Either way, a paper and important structural results were soon published to wide and deserving acclaim.

10.6   ChatGPT and GPT-4

Writing this chapter has been a challenge in that my intent is to bring my history up to the present, but the present keeps changing. Indeed, one of the reviewers for this book complained that this chapter (or the earlier version he saw) already felt three years out of date. This was an exaggeration (it was only six months), but an understandable one, since the start of 2023 saw more publicity and controversy about AI than all the previous sixty-five years combined.

The brouhaha was occasioned by the release of ChatGPT by OpenAI. ChatGPT is a large language model—it is also called GPT-3.5, as it is the successor to GPT-3. However, as opposed to previous LLMs, ChatGPT was made available to the public. Suddenly, it became apparent that this was no longer your grandmother's AI. This program could hold a reasonable conversation or give pretty good responses to requests like a recipe for dairy-free Mac'n Cheese. The first few lines are:

Yes, you can make a delicious dairy-free Mac'n Cheese with
a few simple ingredients. Here is what you will need:

- 1 pound of elbow macaroni

- 2 cups of unsweetened almond milk

- 1/2 cup of nutritional yeast

Your author is not a good enough cook to judge the quality of the recipe,
but I do notice the substitution of almond milk for regular milk and
omission of any cheese—both necessary if the result is to be dairy-free.

However, it was possible to get ChatGPT to say things that could
upset people (known as *toxic content*). Most famously, despite OpenAI's
best efforts, a *New York Times* reporter got a closely related model to
start claiming that he did not really love his wife. OpenAI changed the
program to avoid this behavior in the future. It took a long conversation
to manipulate the program into behaving this way, and, subsequently,
such long conversations with ChatGPT were barred. It is also the case
that the program often says things that are simply false. For example,
when asked if there are prime factors of thirty that differ by three, it
gets it wrong. The correct answer is that the prime factors of thirty are
two, three, and five. Two and five differ by three, so the answer is "Yes,
two and five." ChatGPT finds that answer, but also claims that three
and seven also are answers. This problem is called "hallucination." (We
saw an earlier example of this when GPT-2 described a unicorn as "a
three-horned beast.")

Explaining how ChatGPT differs from GPT-3 is complicated. Chat-
GPT starts with GPT-3 as its base and, thus, can be correctly described
as an LLM. However, it moves beyond that description in that it also uses
reinforcement learning (see section 9.3). Note that LLMs by themselves
have little means to consider the entire textual response as a whole. In
reinforcement learning, the interesting point is that we must necessarily
wait until we get a reward and then try to propagate the reward signal
back to earlier moves so we can make better choices from beginning to
end. For ChatGPT, the creators got humans to label alternative, complete GPT-3 responses with their preferences. Applying reinforcement
learning to that data produced more integrated responses from Chat-
GPT [77].

GPT-4 followed a few months later. Besides being much more accurate than ChatGPT, it has one significant addition—it can accept images
as part of the prompt. Keeping with our culinary examples, I was particularly struck by its good response to the prompt "What can I make
with these ingredients?" followed by a picture of an open refrigerator.

Unfortunately, OpenAI has been completely closed-mouth about GPT-4, and has published nothing about how it works. Obviously, it is an LLM, but its number of parameters, the nature of its training data, and other aspects are all secret. (A technical article on the project logs in at over 100 pages, but it gives data only on GPT-4's accomplishments and nothing about how it works [76]).

One final note: besides writing, another occupation that requires putting down one symbol after another is programming, and both Chat-GPT and GPT-4 have been trained on not just language texts but also computer programs. GPT-4, in particular, is pretty good. I do not think programmers are in danger of mass layoffs, but it will change the nature of the job. I know it is already changing college-level courses on the topic.

11

AI, Present and Future (2023)

No writing on the future of anything is safe without a reference to the aphorism, "Prediction is difficult, especially about the future." I gather it is an old Danish saying, but I enjoy believing the (false) attribution to the baseball player Yogi Berra. (Berra got credit for this, no doubt, because he has come up with so many of these types of sayings, like "It's déjà vu all over again.") The perils of prediction, thus, make it tempting in an AI history to stop before considering its future, but "what's past is prologue," as Shakespeare is wont to say, and a prologue is a prologue to the future.

So, what does the future of AI look like? Meteorologists are careful to distinguish weather from climate. In AI, we have a succession of climate changes from AI summers to AI winters and back again. In each case, the problem was "irrational exuberance," to use a phrase popularized by Alan Greenspan, a former head of the US Federal Reserve, though he was talking about the stock market.

To some degree, this is not surprising. If researchers were not hopeful that recent ideas would allow for significant immediate progress, I doubt we could dedicate ourselves to the effort. Furthermore, if you can't get enthusiastic about your work, it is unlikely that others will do it for you.

Usually AI summers go unmentioned until the temperature drops, at which point the field looks around, trying to figure out what went wrong. (But a good account of the AI summers over the years is due to Henry Kautz [50]). The first AI winter, back in 1973, is associated with the "Lighthill Report." Sir James Lighthill, an applied mathematician at

Cambridge University, was asked by the British government for a report on AI research. His report was quite blunt and quite negative:

> In no part of the field have the discoveries made so far produced the major impact that was then promised [57].

The only person and research to be spared was Terry Winograd and his work on language processing. SHRDLU impressed even Lighthill, but he classified it as an exception and did not think it would have any long-term impact. Lighthill recommended that the British government stop funding AI research, and it did. In the United States, the Defense Department stopped funding AI as well. It is hypothesized that the rise of expert system research in the 1970s (chapter 2) was a reaction to Lighthill, though, as we have seen, it also led to naught.

The second AI winter had similar overoptimistic causes but different effects. Back in 1973, there was virtually no private (as opposed to government) funding for AI. By the time the second AI winter came around in 1988, the landscape was quite different. This was end of the expert system period of AI, so companies were becoming aware of a new technology on the block and many hoped that it could solve some some pressing problems for them. But again, things did not work out. R1 was a success for Digital Equipment Corporation, but the successes were few and far between. R1 figures very prominently in most AI histories because it is a well-documented success from that period, but I know of no other, and if you claimed that it was the only one, I could not gainsay you.

I found no documentation of the following, but I remember being told that shortly after R1's creation, DEC found they no longer needed the program. The *real* problem was that nobody at DEC had ever sat down and tried to systematize the process of configuring VAX computers. Once they had to do so to create R1, they realized a few simple rules enabled a person to get it right. Even if this is true, R1 was a success in a way, but somehow, without a program, this success looks less impressive.

In retrospect, the buildup to this second AI winter had something almost feverish about it. One indelible memory for me is of the spectacular meltdown of a startup I won't name. I knew several of the researchers there, and despite their stupidity, in this instance, I felt (and still feel) sorry for them, although some might say I am too tenderhearted. The company was preparing to demo their technology at a major AI conference and worried that something might go wrong, even though they had a tested script for a demo that would show off the technology to its best advantage. They knew the program could perform the tasks they were going to ask of it. As I said, with consummate stupidity, they put in a backup, so that if the program somehow did not come up with the right answer, it could fall back on a list of answers buried somewhere in the code. Well, as the saying goes, it was the perfect storm. Not only did the program think it needed to fall back on the canned answers, but the responses it made were still not the correct ones. When someone in the audience observed that the program was giving the response not to the current input but, rather, to what was scheduled next ... well, it was a mess. As I said, I knew these folks, and I am a sufficiently flawed person that I can think, "There but by the grace of God go I." (Or, as the religious skeptic Christopher Hitchens put it in his book, *God Is Not Great* [36], "there by the grace of God go they.")

I find myself conflicted about the AI winters. Most AI researchers disagree with me, but in my estimation, AI has little to show for its first fifty years. The best I can say is what I said in the preface: the ideas were reasonable, and if they did not lead anywhere useful, well, we could not have known that until they were tried.

This is not the majority view, however. Some areas of classical AI— I am thinking here of classical planning and Bayes nets—are making progress sufficient to encourage their practitioners. Furthermore, large language models are impressive, but their abilities have large gaps. Common sense reasoning is a big one, so there are still some other ingredients required. Since a lot of what is missing looks like inference, the thought is that we are going to need to mix in old research ideas. Knowledge representation, classical reasoning under uncertainty, and search will be recycled, and, thus, their current disuse is only temporary [7,56,59]. Of course, how to do this mixing is not obvious. My feeling is that incorporating these old topics into neural network models will prove sufficiently difficult that the solutions will have to be discovered from scratch. References to the old work will be little more than hand waves from current researchers to their predecessors. DeepMind's AlphaFold is the modern incarnation of an expert system, but nobody calls it that, and in my inspection (admittedly quick) of its references, I did not find any to the expert system literature. I could make a similar point about, say, the molecular engineering work of Regina Barzilay and Tommi Jaakkola (along with their chemistry coauthors) [18].

There is more agreement about the permanence of neural nets, large language models, reinforcement learning, and their ilk, at least if you judge by where the community is putting its effort. But there are naysayers. LLMs are particularly subject to criticism. Most spectacularly, the LLM critics Timnit Gebru and Emily Bender called them "stochastic parrots" [4], and Gebru got fired by Google for her trouble.

Well, there is a sense in which LLMs are repeating the language they have been fed, but aren't we all? And while people are wrong to ascribe "real" intelligence to them, I would say they are *reactively intelligent*. They are very good at reacting to prompts, and in many situations, that is all you need. (And as I mentioned in my discussion of DALL-E, I think it *must* have a fairly deep understanding of most of the words in the descriptions it is fed—it certainly has "<whatever> riding a <whatever>" down pat.)

At the opposite end of the spectrum are those who believe that AI is going to be not only successful but so successful that AI agents are going to (a) put us out of work, (b) kill us all off, or (c) destroy the Earth—take your pick. Variations on these themes go under the rubric of the "AI apocalypse" or "the singularity." The general idea is that, at some point—it always seems to be about twenty years off—AI agents become as smart as us. Furthermore, since there is no reason to think that our level of intelligence is the maximum possible, once they are as smart as us, it will not be too long until they are smarter.

An open letter from the Future of Life Institute circulated on the Web in March 2023, calling for a minimum six-month pause on building LLMs more powerful than GPT-4. It went on to add:

> Should we let machines flood our information channels with propaganda and untruth? Should we automate away all the jobs, including the fulfilling ones? Should we develop non-human minds that might eventually outnumber, outsmart, obsolete and replace us? Should we risk loss of control of our civilization? [54]

To add fuel to the fire, it asks, if this is not agreed to voluntarily, that "governments should step in." This is scary stuff—not AI, but the idea that in the twenty-first-century, there are calls to put engineering and scientific researchers in jail.

Personally, I think they have seen one too many *Terminator* movies. The idea of AIs just taking over the world is implausible. People are as they are because evolution favors those who take care to produce many offspring, even at the expense of other people. AI agents are going to be built, not evolved, so unless their builders go to a lot of trouble, AIs will have no such compulsions.

Another prognostication is less violent, but equally dangerous—the *mismatch problem*, also known as the *alignment problem*. AIs will have some goals, the critics point out, if only the ones we give them. A favorite, for instance, is making paper clips, so this is also known as the *paper clip problem*. The worry is that these short-term goals will not coincide with our long-term goals, such as survival. The AIs, with blatant disregard to common sense (which, presumably they do not have), proceed to turn natural resources into paper clips until the end of the world as we know it. This possibility has been put forward most forcefully by Nick Bostrom [6], a philosopher at Oxford, and Stuart Russell [89], a computer scientist at Berkeley. As for me, I tend to side with the cognitive scientist Steven Pinker who said about the AI apocalypse in general:

> It depends on the hypothesis that humans are so gifted that they can design an omniscient and omnipotent AI, yet so moronic that they would give it control of the universe without testing how it works [79].

To which I would only add, "or including an off switch."

The least disastrous of the AI apocalypse scenarios is that AI simply puts us all out of work. Of course, it is true that automation has made many jobs obsolete. The word "sabotage" comes from the French "sabot," a shoe made from a block of wood. Early in the Industrial Revolution, workers would stop machine production of goods by placing their wooden shoes in the machines, thereby wooding up the works. A century or two later, telephone operators lost their jobs when telephone calls no longer required a person to connect one line with another. (And a good thing, too. It was estimated that, at the rate telephone usage was expanding, within a few years the entire population of the United States would have to be drafted into the telephone operator corps.) Today many customer service workers are being replaced by speech recognition systems, to the frustration of us all. (The good news is speech recognition is going to get better when large language models make their way into industrial use.)

But putting us all out of work? Think about the workers you have seen today. With the possible exception of the truck drivers, I do not think any of them are going to be out of a job soon. (And even the truck drivers are safe for a while. My guess is that full driver automation is still twenty years off. The safety problems are just too great.) I have not been out much, and, besides truck drivers, the only job I observed today is cook at the Indian restaurant where I ordered takeout. The ingredients that a cook needs to manage and the number of cooking techniques he uses are too numerous for a simple technological solution like an automatic coffee vending machine, and a humanoid robot to do the cooking is far in the future. Just getting a robot to pick up a spatula and turn over a chicken cutlet is way beyond anything outside of a research lab. If I had seen a brick layer, the verdict would have been the same, not to mention a butcher, a baker, and, I dare say, a candle-stick maker.

I am a big fan of the Boston Dynamics robots dancing to "Do You Love Me" (figure 11.1). But the manipulators on the robots are quite primitive compared to their locomotion. Boston Dynamics's industrial robot product for loading and unloading trucks only handles boxes, and it does not have grippers but, rather, uses suction cups. I am sure

Figure 11.1: Boston Dynamics robots dancing to "Do You Love Me"

Boston Dynamics is a great company, and I would love to work there, but they are not profitable, and the economic commentators online do not even see a "path to profitability." The commentators are clearly not expecting a rash of automation anytime soon. The physical aspects of robotics is *hard*—in my estimation, much harder than the rest of AI, and to a large degree separate. I had a reason for not including robotics in this history.

More generally, the US government tracks "worker productivity." Roughly, this statistic refers to the total goods and services (the gross domestic product) divided by the number of workers. It is bothersome that this measure has not increased since the 1990s. It is primarily through automation that statistics such as this improve, so, for what it's worth, I am more concerned that AI is going to do too little for the economy, not too much. Mismatches between companies' and governmental expectations for AI and the realities we face have been the source of AI winters in the past, and I believe there is better-than-even odds that they will be again.

But that is speculation. Right now, we are in a prolonged AI summer, and I expect it to continue for a good while. Furthermore, I expect any downturns will be mild, because our fundamentals are sound. Deep learning is not hype. If you are an AI researcher and your topic is, say, a game of perfect information, you will turn to deep learning. And you will do so for games of chance and multiplayer games, understanding chemical engineering and understanding biology, machine translation and chatbots, computer object recognition and computer scene generation. (And remember my programming job at Argonne Lab—deep learning is now the method of choice for distinguishing significant interactions from those high-energy collisions in which nothing important happens [48].)

This cannot be coincidence. Rather, deep learning is a major accomplishment of late twentieth/early twenty-first-century science. The only things that measure up are the discovery (in physics) of dark matter and dark energy and (in biology) the sequencing of the human genome. Large language models in particular are telling us something profound about the nature of understanding in both computers and people, even if we don't understand yet *what* it is telling us.

Researchers are still pulling interesting and unexpected rabbits out of the deep reinforcement learning hat. As I am writing this, DeepMind has made another interesting discovery, this time in mathematics. The program is AlphaTensor, and it improves on results in matrix multiplication [26]. I do not understand the results well enough to say how important it is, but using reinforcement learning NNs to solve purely mathematical problems is certainly unexpected to my eyes.

In short, when my editor asked if my AI history is in service of a particular thesis, I said it is a full-throated defense/advocacy of deep learning and its consequences. I believe I can rest my case.

For many years, I was somewhat pessimistic about AI. It has always been an interesting field, and I am glad I learned of it more than a half-century ago, but I did not expect that we would see a true artificial intelligence for hundreds of years. I never wrote this down, but when people asked me, I would venture 500 years or so. However, the last fifteen years of progress has changed all that—intelligent computers will not be in my lifetime, but perhaps my grandson will see them.

All in all, I like to think that Turing would be proud.

AI Time Line

This section gives some dates for programs and events that have been discussed herein. I have included it to help place events relative to each other as my presentation is not always chronological. It is not particularly intended as a distillation of *the* important AI events. (For that matter, I suppose, this book is not so intended, either!)

1950   Publication of "Computing Machinery and Intelligence"
1952   Checkers-playing program
1955   Early work on what was to become reinforcement learning
1956   Logic Theorist
1956   Term AI is coined at Dartmouth Conference
1957   First full-board chess program
1958   Perceptron learning algorithm
1961   CYK parsing algorithm
1965   Early research on the cat's visual cortex
1966   Hidden Markov models
1966   Mac Hack, first tournament-level chess program
1967   Viterbi algorithm
1967   Publication of *Perceptrons*
1968   SEE program and others on block recognition
1969   Heuristic DENDRAL program
1969   SHRDLU program
1971   STRIPS planner
1973   Lighthill report—first AI winter
1973   Chess 4.0 starts full-width-search chess-playing programs
1975   Nonlinear planning
1975   MYCIN program
1980   Nonmonotonic logics
1980   R1 program
1985   ChipTest, first direct precursor to Deep Blue
1986   Backpropagation
1986   *Parallel Distributed Processing* volumes

1988   Bayesian networks
1988   Second AI winter
1988   Deep Thought, direct precursor to Deep Blue
1989   Deep learning for digit recognition
1990   IBM statistical machine translation
1990   Renewed interest in reinforcement learning
1992   TD-Gammon plays strong backgammon with reinforcement
           learning
1994   Penn treebank
1995   Graphical models for language processing
1996   Temporal difference error/dopamine correspondence
           hypothesis
1997   Deep Blue beats world chess champion, Garry Kasparov
1997   Long short-term memory
1998   Planning Domain Definition Language
1999   SIFT features for image processing
2005   PASCAL Visual Object Classes Challenge
2006   Google launches machine translation service using
           statistical MT
2006   HOG features for image processing
2007   Nvidia releases CUDA—GPU use for general purpose computing
2010   ImageNet Large Scale Visual Recognition Challenge
2012   AlexNet NN wins computer vision challenge
2013   Popularization of word embeddings
2013   Variational autoencoders
2016   AlphaGo beats Lee Sedol, third-ranked Go champion
2016   Google Translate begins switching to neural network MT
2017   Transformers
2018   GPT language model (0.1 billion parameters)
2019   GPT-2 language model (1.5 billion parameters)
2020   AlphaFold largely solves the protein structure prediction problem
2020   GPT-3 language model (175 billion parameters)
2021   DALL-E language-to-image neural networks
2022   PaLM language model (540 billion parameters)
2023   ChatGPT and GPT-4

Bibliography

[1] James K. Baker. Trainable grammars for speech recognition. *The Journal of the Acoustical Society of America*, 65(S1):S132–S132, 1979.

[2] Leonard E. Baum and Ted Petrie. Statistical inference for probabilistic functions of finite state Markov chains. *The Annals of Mathematical Statstics*, 37(6):1554–1563, 1966.

[3] Richard Bellman and Robert Kalaba. Dynamic programming and statistical communication theory. *Proceedings of the National Academy of Sciences*, 43(8):749–751, 1957.

[4] Emily M. Bender, Timnit Gebru, Angelina McMillan-Major, and Shmargaret Shmitchell. On the dangers of stochastic parrots: Can language models be too big? In *Proceedings of the 2021 ACM Conference on Fairness, Accountability, and Transparency*, 610–623, 2021.

[5] Alan Bernstein, T. Arbuckle, M. De V. Roberts, and M. A. Belsky. A chess playing program for the IBM 704. In *Proceedings of the May 6–8, 1958, Western Joint Computer Conference: Contrasts in Computers*, 157–159, 1958.

[6] Nick Bostrom. How long before superintelligence? *International Journal of Futures Studies*, 2, 1998.

[7] Ronald J. Brachman and Hector J. Levesque. Toward a new science of common sense. In *Proceedings of the 36th AAAI Conference on Artificial Intelligence*, 12245–12249, 2022.

[8] Noam Brown and Tuomas Sandholm. Superhuman AI for multiplayer poker. *Science*, 365(6456):885–890, 2019.

[9] Peter F. Brown, John Cocke, Stephen A. Della Pietra, Vincent J. Della Pietra, Frederick Jelinek, John Lafferty, Robert L. Mercer,

and Paul S. Roossin. A statistical approach to machine translation. *Computational Linguistics*, 16(2):79–85, 1990.

[10] Tom Brown, Benjamin Mann, Nick Ryder, Melanie Subbiah, Jared D. Kaplan, Prafulla Dhariwal, Arvind Neelakantan, Pranav Shyam, Girish Sastry, Amanda Askell, et al. Language models are few-shot learners. *Advances in Neural Information Processing Systems*, 33:1877–1901, 2020.

[11] Bruce Buchanan, Georgia Sutherland, and Edward A. Feigenbaum. Heuristic DENDRAL: A program for generating explanatory hypotheses. *Organic Chemistry*, 1969.

[12] Bruce G. Buchanan and Edward H. Shortliffe. *Rule-Based Expert Systems: The MYCIN Experiments of the Stanford Heuristic Programming Project*. Addison-Wesley Publishing Company, 1984.

[13] David J. Chalmers et al. Consciousness and its place in nature. *Blackwell Guide to the Philosophy of Mind*, 102–142, 2003.

[14] Eugene Charniak. *Toward a model of children's story comprehension*. PhD thesis, Massachusetts Institute of Technology, 1972.

[15] Eugene Charniak. Bayesian networks without tears. *AI Magazine*, 12(4):50–50, 1991.

[16] Aakanksha Chowdhery, Sharan Narang, Jacob Devlin, Maarten Bosma, Gaurav Mishra, Adam Roberts, Paul Barham, Hyung Won Chung, Charles Sutton, Sebastian Gehrmann, et al. PaLM: Scaling language modeling with pathways. arXiv preprint arXiv:2204.02311, 2022.

[17] Maxwell B. Clowes. On seeing things. *Artificial Intelligence*, 2(1):79–116, 1971.

[18] Connor W. Coley, Wengong Jin, Luke Rogers, Timothy F. Jamison, Tommi S. Jaakkola, William H. Green, Regina Barzilay, and Klavs F. Jensen. A graph-convolutional neural network model for the prediction of chemical reactivity. *Chemical Science*, 10(2):370–377, 2019.

[19] Eli Collins and Zoubin Ghahramani. LaMDA: Our breakthrough conversation technology. *The Keyword*, May 18. 2021. https://blog.google/technology/ai/lamda/.

[20] Michael Collins. Head-driven statistical models for natural language parsing. *Computational Linguistics*, 29(4):589–637, 2003.

[21] Navneet Dalal and Bill Triggs. Histograms of oriented gradients for human detection. In *2005 IEEE Computer Society Conference on Computer Vision and Pattern Recognition (CVPR'05)*, volume 1, 886–893. IEEE, 2005.

[22] Ernest Davis. *Representations of Commonsense Knowledge*. Morgan Kaufmann, 2014.

[23] Prafulla Dhariwal and Alexander Nichol. Diffusion models beat GANs on image synthesis. *Advances in Neural Information Processing Systems*, 34:8780–8794, 2021.

[24] M. Everingham, L. Van Gool, C. K. I. Williams, J. Winn, and A. Zisserman. The Pascal Visual Object Classes (VOC) Challenge. *International Journal of Computer Vision*, 88(2):303–338, June 2010.

[25] Kevin Eykholt, Ivan Evtimov, Earlence Fernandes, Bo Li, Amir Rahmati, Chaowei Xiao, Atul Prakash, Tadayoshi Kohno, and Dawn Song. Robust physical-world attacks on deep learning visual classification. In *Proceedings of the IEEE Conference on Computer Vision and Pattern Recognition*, 1625–1634, 2018.

[26] Alhussein Fawzi, Matej Balog, Aja Huang, Thomas Hubert, Bernardino Romera-Paredes, Mohammadamin Barekatain, Alexander Novikov, Francisco J. R. Ruiz, Julian Schrittwieser, Grzegorz Swirszcz, et al. Discovering faster matrix multiplication algorithms with reinforcement learning. *Nature*, 610(7930):47–53, 2022.

[27] Pedro Felzenszwalb, David McAllester, and Deva Ramanan. A discriminatively trained, multiscale, deformable part model. In *2008 IEEE Conference on Computer Vision and Pattern Recognition*, 1–8. IEEE, 2008.

[28] Richard E. Fikes and Nils J. Nilsson. STRIPS: A new approach to the application of theorem proving to problem solving. *Artificial Intelligence*, 2(3-4):189–208, 1971.

[29] Ross Girshick, Jeff Donahue, Trevor Darrell, and Jitendra Malik. Rich feature hierarchies for accurate object detection and semantic segmentation. In *Proceedings of the IEEE Conference on Computer Vision and Pattern Recognition*, 580–587, 2014.

[30] Ian Goodfellow, Yoshua Bengio, and Aaron Courville. *Deep Learning*. MIT Press, 2016.

[31] Richard D. Greenblatt, Donald E. Eastlake III, and Stephen D. Crocker. The Greenblatt chess program. In *Proceedings of the November 14–16, 1967, Rall Joint Computer Conference*, 801–810, 1967.

[32] Adolfo Guzmán. Decomposition of a visual scene into three-dimensional bodies. In *Proceedings of the December 9–11, 1968, Fall Joint Computer Conference, Part I*, 291–304, 1968.

[33] Patrick J. Hayes. The naive physics manifesto. *Expert Systems in the Microelectronic Age*, 1979.

[34] Patrick J. Hayes. Naive physics I: Ontology for liquids. In *Readings in Qualitative Reasoning about Physical Systems*, 484–502. Morgan Kaufmann, 1990.

[35] Martin Herman and Takeo Kanade. Incremental reconstruction of 3D scenes from multiple, complex images. *Artificial Intelligence*, 30(3):289–341, 1986.

[36] Christopher Hitchens. *God Is Not Great: How Religion Poisons Everything*. McClelland & Stewart, 2008.

[37] Thomas Hobbes and Marshall Missner. *Thomas Hobbes: Leviathan (Longman Library of Primary Sources in Philosophy)*. Routledge, 2016.

[38] Sepp Hochreiter and Jürgen Schmidhuber. Long short-term memory. *Neural Computation*, 9(8):1735–1780, 1997.

[39] Feng-Hsiung Hsu. *Large-scale parallelization of alpha-beta search: An algorithmic and architectural study with computer chess*. PhD thesis. Carnegie Mellon University, 1989.

[40] Feng-hsiung Hsu, Murray S. Campbell, and A. Joseph Hoane Jr. Deep blue system overview. In *Proceedings of the 9th International Conference on Supercomputing*, 240–244, 1995.

[41] David H. Hubel and Torsten N. Wiesel. Receptive fields, binocular interaction and functional architecture in the cat's visual cortex. *The Journal of Physiology*, 160(1):106, 1962.

[42] David A. Huffman. Impossible object as nonsense sentences. *Machine Intelligence*, 6:295–324, 1971.

[43] Frederick Jelinek. Continuous speech recognition by statistical methods. *Proceedings of the IEEE*, 64(4):532–556, 1976.

[44] Béla Julesz. *Foundations of Cyclopean Perception.* University of Chicago Press, 1971.

[45] John Jumper, Richard Evans, Alexander Pritzel, Tim Green, Michael Figurnov, Olaf Ronneberger, Kathryn Tunyasuvunakool, Russ Bates, Augustin Žídek, Anna Potapenko, et al. Highly accurate protein structure prediction with AlphaFold. *Nature,* 596(7873):583–589, 2021.

[46] Leslie Pack Kaelbling, Michael L. Littman, and Andrew W. Moore. Reinforcement learning: A survey. *Journal of Artificial Intelligence Research,* 4:237–285, 1996.

[47] Daniel Kahneman, Stewart Paul Slovic, Paul Slovic, and Amos Tversky. *Judgment under Uncertainty: Heuristics and Biases.* Cambridge University Press, 1982.

[48] Georgia Karagiorgi, Gregor Kasieczka, Scott Kravitz, Benjamin Nachman, and David Shih. Machine learning in the search for new fundamental physics. *Nature Reviews Physics,* 4(6):399–412, 2022.

[49] Tadao Kasami. An efficient recognition and syntax-analysis algorithm for context-free languages. *Coordinated Science Laboratory Report no. R-257,* 1966.

[50] Henry Kautz. The third AI summer: AAAI Robert S. Engelmore Memorial Lecture. *AI Magazine,* 43(1):93–104, 2022.

[51] Diederik P. Kingma and Max Welling. Auto-encoding variational Bayes. arXiv preprint arXiv:1312.6114, 2013.

[52] Alex Krizhevsky, Ilya Sutskever, and Geoffrey E. Hinton. Imagenet classification with deep convolutional neural networks. *Advances in Neural Information Processing Systems,* 25, 2012.

[53] Yann LeCun, Bernhard Boser, John S. Denker, Donnie Henderson, Richard E. Howard, Wayne Hubbard, and Lawrence D. Jackel. Backpropagation applied to handwritten zip code recognition. *Neural Computation,* 1(4):541–551, 1989.

[54] FOL Open Letters. Pause giant AI experiments: An open letter. *Future of Life Institution.* https://futureoflife.org/open-letter /pause-giant-ai-experiments, 2023.

[55] Hector Levesque, Ernest Davis, and Leora Morgenstern. The Winograd Schema Challenge. In *Thirteenth International Conference on the Principles of Knowledge Representation and Reasoning,* 2012.

[56] Hector J. Levesque. *Common Sense, the Turing Test, and the Quest for Real AI.* MIT Press, 2017.

[57] James Lighthill. *Artificial Intelligence: A General Survey.* Science Research Council, 1973.

[58] David G. Lowe. Distinctive image features from scale-invariant keypoints. *International Journal of Computer Vision*, 60(2):91–110, 2004.

[59] Gary Marcus and Ernest Davis. *Rebooting AI: Building Artificial Intelligence We Can Trust.* Vintage, 2019.

[60] Gary Marcus, Ernest Davis, and Scott Aaronson. A very preliminary analysis of DALL-E 2. arXiv preprint arXiv:2204.13807, 2022.

[61] Mitch Marcus, Grace Kim, Mary Ann Marcinkiewicz, Robert MacIntyre, Ann Bies, Mark Ferguson, Karen Katz, and Britta Schasberger. The Penn treebank: Annotating predicate argument structure. In *Human Language Technology: Proceedings of a Workshop Held at Plainsboro, New Jersey, March 8–11*, 1994.

[62] David Marr and Tomaso Poggio. A computational theory of human stereo vision. *Proceedings of the Royal Society of London. Series B. Biological Sciences*, 204(1156):301–328, 1979.

[63] John McCarthy. Circumscription—a form of non-monotonic reasoning. *Artificial Intelligence*, 13(1-2):27–39, 1980.

[64] John McCarthy, Marvin L. Minsky, Nathaniel Rochester, and Claude E. Shannon. A proposal for the Dartmouth summer research project on artificial intelligence, August 31, 1955. *AI Magazine*, 27(4):12–12, 2006.

[65] Pamela McCorduck, CLI CFE. *Machines Who Think: A Personal Inquiry into the History and Prospects of Artificial Intelligence.* CRC Press, 2004.

[66] Warren S. McCulloch and Walter Pitts. A logical calculus of the ideas immanent in nervous activity. *The Bulletin of Mathematical Biophysics*, 5(4):115–133, 1943.

[67] Drew McDermott and Jon Doyle. Non-monotonic logic I. *Artificial Intelligence*, 13(1-2):41–72, 1980.

[68] John McDermott. R1: A rule-based configurer of computer systems. *Artificial Intelligence*, 19(1):39–88, 1982.

[69] Tomas Mikolov, Ilya Sutskever, Kai Chen, Greg S. Corrado, and Jeff Dean. Distributed representations of words and phrases and their compositionality. *Advances in Neural Information Processing Systems*, 26, 2013.

[70] Marvin Minsky. Steps toward artificial intelligence. *Proceedings of the IRE*, 49(1):8–30, 1961.

[71] Marvin Minsky. Artificial intelligence. *Scientific American*, 215(3), September 1966.

[72] Marvin Minsky. *A Framework for Representing Knowledge*. MIT, Cambridge, 1974.

[73] Marvin Minsky and Seymour A. Papert. *Perceptrons, Reissue of the 1988 Expanded Edition with a New Foreword by Léon Bottou: An Introduction to Computational Geometry*. MIT Press, 2017.

[74] P. Read Montague, Peter Dayan, and Terrence J. Sejnowski. A framework for mesencephalic dopamine systems based on predictive Hebbian learning. *Journal of Neuroscience*, 16(5):1936–1947, 1996.

[75] Allen Newell, John Clifford Shaw, and Herbert A. Simon. Empirical explorations of the logic theory machine: A case study in heuristic. In *Papers Presented at the February 26–28, 1957, Western Joint Computer Conference: Techniques for Reliability*, 218–230, 1957.

[76] OpenAI. *GPT-4 technical report*, 2023. https://arxiv.org/abs/2303.08774.

[77] Long Ouyang, Jeffrey Wu, Xu Jiang, Diogo Almeida, Carroll Wainwright, Pamela Mishkin, Chong Zhang, Sandhini Agarwal, Katarina Slama, Alex Ray, et al. Training language models to follow instructions with human feedback. *Advances in Neural Information Processing Systems*, 35:27730–27744, 2022.

[78] Judea Pearl. *Probabilistic Reasoning in Intelligent Systems: Networks of Plausible Inference*. Morgan Kaufmann, 1988.

[79] Steven Pinker. *Enlightenment Now: The Case for Reason, Science, Humanism, and Progress*. Penguin UK, 2018.

[80] Alec Radford, Jeffrey Wu, Rewon Child, David Luan, Dario Amodei, Ilya Sutskever, et al. Language models are unsupervised multitask learners. *OpenAI*, 1(8):9, 2019.

[81] Aditya Ramesh, Prafulla Dhariwal, Alex Nichol, Casey Chu, and Mark Chen. Hierarchical text-conditional image generation with clip latents. arXiv preprint arXiv:2204.06125, 2022.

[82] Raymond Reiter. A logic for default reasoning. *Artificial Intelligence*, 13(1–2):81–132, 1980.

[83] Ida Rhodes. A new approach to the mechanical syntactic analysis of Russian. *Mech. Transl. Comput. Linguistics*, 6:33–50, 1961.

[84] Frank Rosenblatt. The perceptron: a probabilistic model for information storage and organization in the brain. *Psychological Review*, 65(6):386, 1958.

[85] David E. Rumelhart, Geoffrey E. Hinton, and Ronald J. Williams. Learning representations by back-propagating errors. *Nature*, 323(6088):533–536, 1986.

[86] David E. Rumelhart, James L. McClelland, PDP Research Group, et al. *Parallel Distributed Processing*, volume 1. IEEE New York, 1988.

[87] Olga Russakovsky, Jia Deng, Hao Su, Jonathan Krause, Sanjeev Satheesh, Sean Ma, Zhiheng Huang, Andrej Karpathy, Aditya Khosla, Michael Bernstein, et al. ImageNet Large Scale Visual Recognition Challenge. *International Journal of Computer Vision*, 115(3):211–252, 2015.

[88] S. J. Russell and Peter Norvig. *Artificial Intelligence: A Modern Approach*, 4th, Global ed. Prentice Hall, 2022.

[89] Stuart Russell. *Human Compatible: Artificial Intelligence and the Problem of Control*. Penguin, 2019.

[90] Earl D. Sacerdoti. The nonlinear nature of plans. Technical report. Stanford Research Institute, 1975.

[91] Arthur L. Samuel. Programming computers to play games. In *Advances in Computers*, volume 1, 165–192. Elsevier, 1960.

[92] Roger C. Schank and Robert P. Abelson. *Scripts, Plans, Goals, and Understanding: An Inquiry into Human Knowledge Structures*. Psychology Press, 2013.

[93] Andrew W. Senior, Richard Evans, John Jumper, James Kirkpatrick, Laurent Sifre, Tim Green, Chongli Qin, Augustin Žídek, Alexander W. R. Nelson, Alex Bridgland, et al. Improved protein

structure prediction using potentials from deep learning. *Nature*, 577(7792):706–710, 2020.

[94] Rico Sennrich, Barry Haddow, and Alexandra Birch. Neural machine translation of rare words with subword units. arXiv preprint arXiv:1508.07909, 2015.

[95] Glenn Shafer. Dempster–Shafer theory. *Encyclopedia of Artificial Intelligence*, 1:330–331, 1992.

[96] Edward H. Shortliffe and Bruce G. Buchanan. A model of inexact reasoning in medicine. *Mathematical Biosciences*, 23(3-4):351–379, 1975.

[97] David Silver, Aja Huang, Chris J. Maddison, Arthur Guez, Laurent Sifre, George Van Den Driessche, Julian Schrittwieser, Ioannis Antonoglou, Veda Panneershelvam, Marc Lanctot, et al. Mastering the game of Go with deep neural networks and tree search. *Nature*, 529(7587):484–489, 2016.

[98] David Silver, Julian Schrittwieser, Karen Simonyan, Ioannis Antonoglou, Aja Huang, Arthur Guez, Thomas Hubert, Lucas Baker, Matthew Lai, Adrian Bolton, et al. Mastering the game of Go without human knowledge. *Nature*, 550(7676):354–359, 2017.

[99] David J. Slate and Lawrence R. Atkin. Chess 4.5—the Northwestern University chess program. In *Chess Skill in Man and Machine*, 82–118. Springer, 1983.

[100] Richard S. Sutton and Andrew G. Barto. *Reinforcement Learning: An Introduction.* MIT Press, 2018.

[101] Christian Szegedy, Wojciech Zaremba, Ilya Sutskever, Joan Bruna, Dumitru Erhan, Ian Goodfellow, and Rob Fergus. Intriguing properties of neural networks. arXiv preprint arXiv:1312.6199, 2013.

[102] Gerald Tesauro. Temporal difference learning and TD-Gammon. *Communications of the ACM*, 38(3):58–68, 1995.

[103] Alan M. Turing. Computing machinery and intelligence. In *Parsing the Turing Test*, 23–65. Springer, 2009.

[104] Ashish Vaswani, Noam Shazeer, Niki Parmar, Jakob Uszkoreit, Llion Jones, Aidan N. Gomez, Łukasz Kaiser, and Illia Polosukhin. Attention is all you need. *Advances in Neural Information Processing Systems*, 30, 2017.

[105] Andrew Viterbi. Error bounds for convolutional codes and an asymptotically optimum decoding algorithm. *IEEE Transactions on Information Theory*, 13(2):260–269, 1967.

[106] David Waltz. Understanding line drawings of scenes with shadows. AI Lab, Massachutts Institute of Technology, 1975.

[107] Alfred North Whitehead and Bertrand Russell. *Principia Mathematica*, volume 2. Cambridge University Press, 1997.

[108] Henry Widdowson. Firth, 1957, Papers in Linguistics 1934–51. *International Journal of Applied Linguistics*, 17(3):402–413, 2007.

[109] Ronald J. Williams. Simple statistical gradient-following algorithms for connectionist reinforcement learning. *Machine Learning*, 8(3):229–256, 1992.

[110] Terry Winograd. Understanding natural language. *Cognitive Psychology*, 3(1):1–191, 1972.

[111] Terry Winograd and Fernando Flores. *Understanding Computers and Cognition: A New Foundation for Design*. Intellect Books, 1986.

[112] Patrick H. Winston. Learning structural descriptions from examples. AI Lab, Massachusetts Institute of Technology, 1970.

[113] Blaise Agüera y Arcas. Do large language models understand us? *Daedalus*, 151(2):183–197, 2022.

[114] Lotfi Asker Zadeh. Fuzzy sets as a basis for a theory of possibility. *Fuzzy Sets and Systems*, 1(1):3–28, 1978.

Index